INTERPRETIVE TECHNIQUES FOR MICROSTRUCTURAL ANALYSIS

INTERPRETIVE TECHNIQUES FOR FOR MICROSTRUCTURAL ANALYSIS

Edited by

James L. McCall

Battelle-Columbus Laboratories
Columbus, Ohio

and

P. M. French

Westinghouse Electric Corporation
Cheswick, Pennsylvania

PLENUM PRESS · NEW YORK AND LONDON

Library of Congress Cataloging in Publication Data

Main entry under title:

Interpretive techniques for microstructural analysis.

"Proceedings of a symposium . . . held in Minneapolis, Minnesota, June 29-30, 1975."
Includes index.
1. Metallography—Congresses. I. McCall, James L. II. French, Peter Michael, 1935-

TN689.2.I63	669'.95	77-2333

ISBN-13: 978-1-4684-2372-3   e-ISBN-13: 978-1-4684-2370-9
DOI: 10.1007/978-1-4684-2370-9

Proceedings of a symposium on Interpretive Techniques for Microstructural Analysis
held in Minneapolis, Minnesota, June 29—30, 1975

© 1977 Plenum Press, New York
A Division of Plenum Publishing Corporation
227 West 17th Street, New York, N.Y. 10011
Softcover reprint of the hardcover 1st edition 1977

PREFACE

In recent years microstructural analysis has been a rapidly changing field of scientific endeavor. No longer are the efforts of the microstructural analysts (sometimes referred to as metallographers, materialographers, ceramographers, and similar designations) limited to the tasks of polishing, etching, and photographing specimens of materials. The performance demanded of materials used for many current applications requires much more complete characterizations than were possible only a scant few years ago. Although the individuals who have been expected to develop new and improved techniques to permit these required characterizations have been severely challenged, in large part they have met the challenge.

In view of the many new developments in the field of microstructural analysis and recognizing the requirements to communicate these developments to the wide audience that might make use of them, the American Society for Metals and the International Metallographic Society joined forces to co-sponsor a symposium that was intended to bring participants and attendees up to date on the subject "Interpretive Techniques for Microstructural Analysis". This symposium was held in Minneapolis, Minnesota, USA, June 29 and 30, 1975. It followed two earlier symposia co-sponsored by the same two societies on other subjects of current interest to the metallographic community, Microstructural Analysis — Tools and Techniques, 1972, and Metallographic Specimen Preparation — Optical and Electron Microscopy, 1973.

The interest shown in the recent symposium encouraged us to publish the presented papers in this volume. We felt these proceedings would serve as a useful reference for all those involved in analyzing microstructures of materials, either on a practical or a theoretical level.

A symposium of the type that was conducted would not have been so successful or, for that matter, even possible, without the combined efforts of many individuals. But special thanks are owed to Mr. J.M. Hoegfeldt, General Chairman of the 1975 International Metallographic Convention of which this symposium was a part. The cooperation of both co-sponsoring societies was assured through several individuals, most directly, Mr. Oren Huber of the American Society for Metals and Dr. E.D. Albrecht of the International Metallographic Society. Finally, we thank Connie McCall for her efforts in putting all the various contributions into a uniform style and in typing and laying them out into camera-ready copy.

We hope a useful document has resulted from the combined efforts of these and many more unnamed individuals.

James L. McCall
Battelle—Columbus Laboratories

P.M. French
Westinghouse Electric Corporation

CONTENTS

INTERPRETIVE TECHNIQUES FOR MICROSTRUCTURAL ANALYSIS

BASIC PHOTOGRAPHIC OPTICS FOR THE METALLURGIST

M. D. ADAMS *

INTRODUCTION

It seems that few people study microscopy any more. Metallurgists, like biologists and chemists, want to see a structure in order to correlate it with some property. Preparation of the specimen may be considered the real science and it is usually the most time consuming part of the study. Use of the microscope is obvious and the application is so revealing, even when optimum conditions are not met, that few feel the call to study this tool independently of their immediate requirement. It is seldom realized even by those who use microscopes extensively that the equipment for light microscopy has been developed almost to the limit of theoretical possibility. It also is seldom realized what a remarkable performance can be achieved when a system is used under optimum conditions.

In order to explain your interpretations to others or to instruct them, it is necessary to be able to produce high quality, magnified, and portable images. It is reasonably safe to say that all metallurgical microscopes have, or can be fitted with, photographic accessories. Polaroid film gives us instant access to a print. If it is not right, change something and take another. A photographic record is readily available. Yet, when a special requirement arises, if the ultimate purpose is a color slide, or if some feature is to be shown which is remarkable because of requirements of lighting, resolution, or of depth of focus, then some appreciation of photo-optical principles and limitations can be most helpful.

There are three major topics involved in photomicrography; light, optics, and film. In this paper the intent is to present some of the relationships among these topics, some limitations faced because of physical laws or equipment, and perhaps help with some practical applications.

LIGHT

Even though the announced topic is photographic optics, the understanding of the nature and behavior of light is so vital that the discussion will begin with these topics.

Light has been considered in terms of its color temperature and intensity. This specifies the quality for most purposes. Traditionally, the metallographer used a carbon

* Packer Engineering Associates, Naperville, Illinois 60540 USA

1

arc with a green filter for critical photomicrography. It is a combination that provides near optimum theoretical and practical conditions for good photographs. The carbon arc provides high intensity light, microscope lenses give optimum resolution in mono-chromatic light of shorter wavelengths, and the traditional films are blue-green sensitive orthochromatic plates.

Carbon arcs are rare now, the usual sources are tungsten filaments in conventional bulbs or in halogen containing quartz envelopes and high pressure xenon arcs.

Emission of light from a hot source is described by the black body laws, to some extent on the material being heated, but mainly by the temperature to which it is raised. Thus, if a tungsten filament is heated by resistance, the spectrum of light emission changes as the filament gets hotter. Radiation begins in the lower energy, long wavelength region of infrared where it can be felt before it is seen, and proceeds into the shorter wavelengths as current increases. Long before a temperature is achieved which might produce light simulating sunlight, the tungsten filament melts.

In order to achieve a higher color temperature, it is necessary to substitute a filament which is thermally stable in place of the tungsten wire. At the temperature of sunlight, considered a standard at about 6000°K, the stable form of matter is plasma. Electric arcs using carbon, rare gases, or metal plasmas can provide good high temperature light sources. Some spectral curves [1] for light sources are shown in Fig. 1.

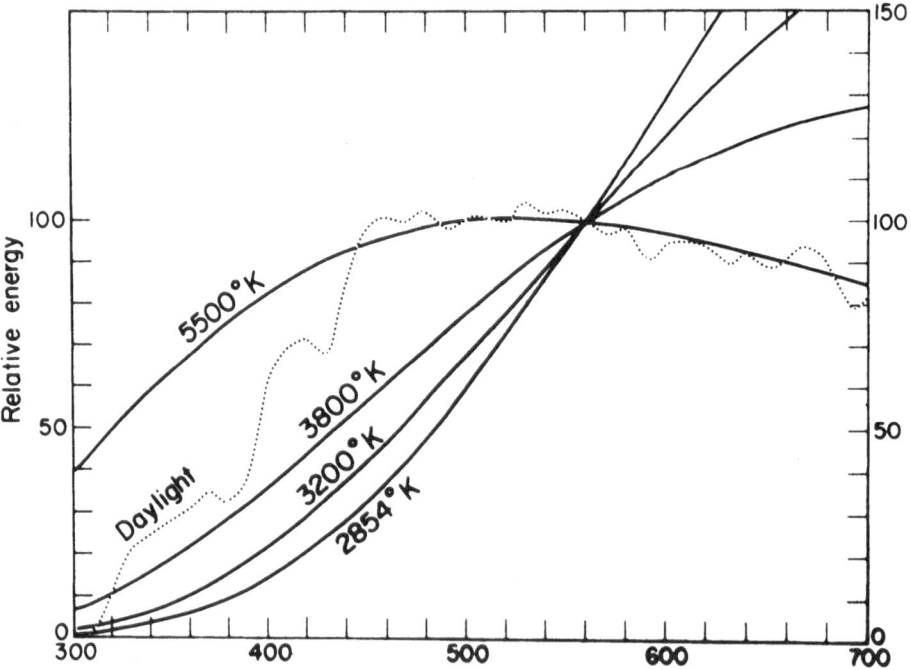

Fig. 1. Relative Spectral Quality of Light of Differing Color Temperature [1].

Modification of light to remove spectral peaks, lower intensity, shift color temperature, or provide a nearly monochromatic wavelength can be done with filters [1,2,3]. According to the application, filters can be classified into four basic types, with perhaps a few subtypes:

1. Neutral-density
2. Cut-off
3. Band-pass or band-absorption
4. Color-compensating

Neutral-density filters, Fig. 2, are used simply for the purpose of lowering intensity without altering the color temperature. Ratings are given in density units,

$$D = -\log_{10}T$$

where T represents percent transmittance. Some typical values are:

D	T
0.3	50%
1.0	10%
2.0	1%
3.0	0.1%

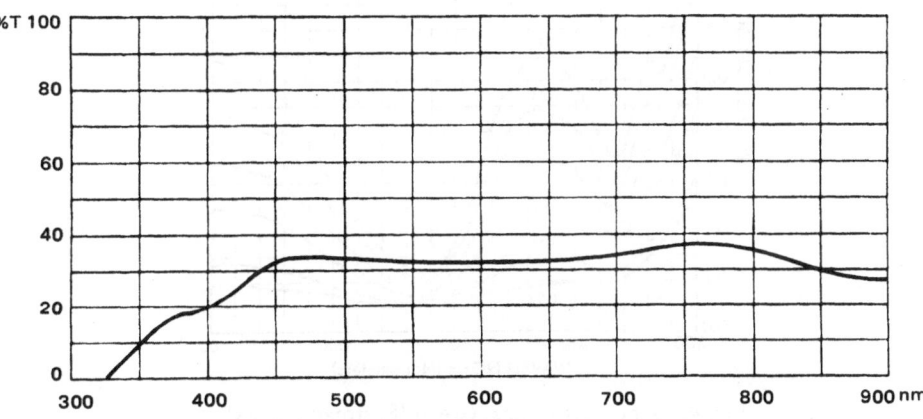

Fig. 2. Neutral Density Filter (Wild Heerbrugg)

Cut-off filters effectively absorb all wavelengths below or above a specified wavelength, Fig. 3. These may be sharp or gradual depending upon the slope of the plot of wavelength vs T. Ultraviolet and infrared eliminating filters belong to this group as well as some color modifying types.

Band-pass filters will absorb wavelengths both shorter and longer than a certain band. A green filter, for instance, must absorb blue and red, Fig. 4. These bands may be quite narrow or include a large fraction of the visible spectrum. As the reverse

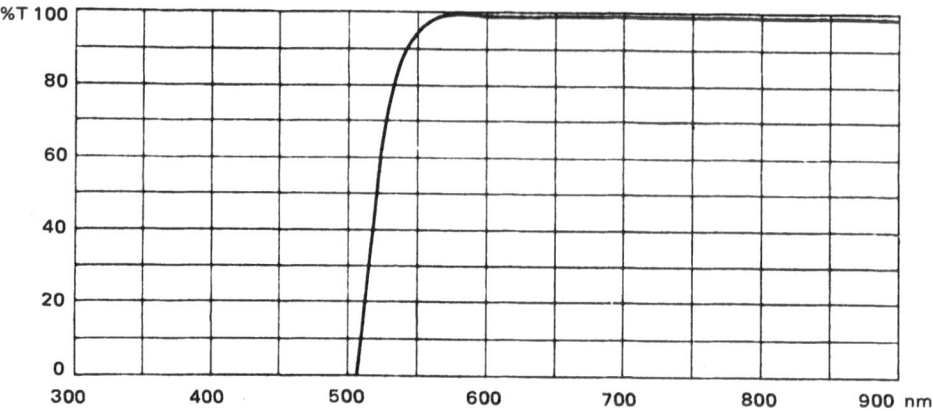

Fig. 3. Cut-off Filter to Absorb Blue Light (Wild Heerbrugg)

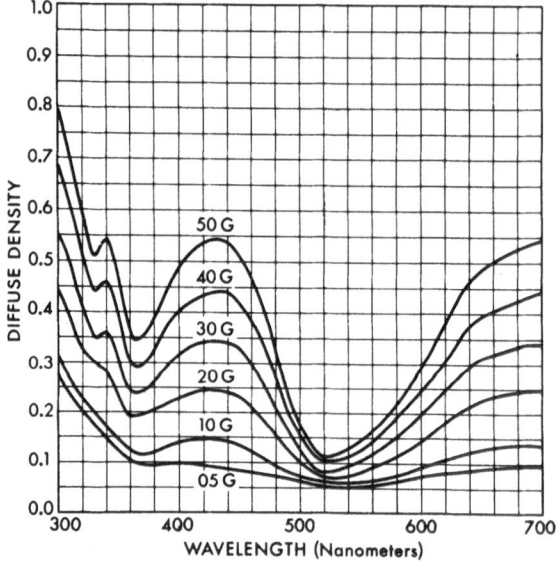

Fig. 4. Band Pass Filters, Green (Eastman Kodak)

case, band-absorbing filters, such as those passing magenta, Fig. 5, absorb in the central wavelengths of the spectrum but pass blue and red.

Color-compensating filters include a number of subtypes for special purposes such as color printing and photography in light of the wrong color balance for the film being used. All are for the purpose of modifying the shape of the spectrum of a light source to conform to the color temperature and the emulsion type, Fig. 6.

The conventional filters are available in glass or dyed gelatin and in general are not precise in wavelength selectivity. Almost as a separate type, interference filters offer a method for obtaining virtually monochromatic band pass capability. These consist

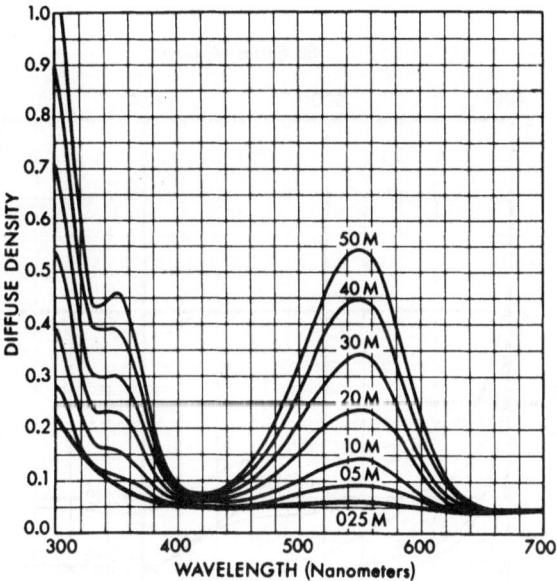

Fig. 5. Magenta Band Absorption Filters (Eastman Kodak)

of multiple thin layers vacuum deposited on a glass substrate in precise thicknesses of
the order of one fourth or one half wavelength. Nearly total reflection or approximately
50% transmission of specific wavelengths can be achieved, Fig. 7, by selection of materials and thickness of the deposited layers. Standard filters are offered by manufacturers
[4] or it is possible to specify a special type of filter and have it produced.

More specific data on filters are available from manufacturers. For the conventional
type, possibly the most readily available reference is the Eastman Kodak Publication B-3
on filters [3]. Spectra are published for each filter in terms of percent transmission
from near UV through the near IR. Using these tables, it is feasible to devise combinations of light source and one or multiple filters which will provide light of a quality which
may be required. As a cautionary note, it is always better to start with a light as close
as possible to what is needed to avoid the losses involved in multiple filters.

OPTICS

Selection of optical components for a given application is a major part of photomicrography. Optimum application of the optics is the other part and is where most of us fall
short. For intelligent use it is necessary to develop some appreciation of the performance
capabilities and limitations of the equipment and match these to the job requirements.

Photo-optics includes getting light of the right quality, intensity, and distribution onto
the specimen; magnifying an image with the proper contrast and resolution; and projecting
it onto the film which is most appropriate for the task. It is necessary to take into account

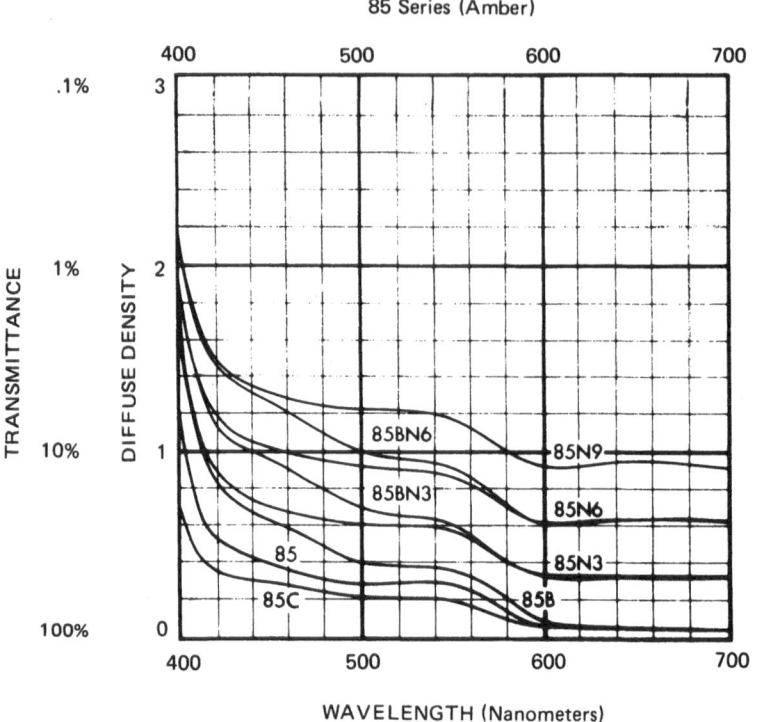

Fig. 6. Color Compensating Filters (Eastman Kodak)

the light quality, optical abberations, depth of field, exposure, resolution, and characteristics of film, all in addition to the metallurgical concerns with sample selection and preparation.

The flexibility and utility of the photomacrographic approach is often overlooked by the metallographer [5,6,7]. This perhaps, provides a good introduction to the general topic of photo-optics. As it is usually defined, it is a region in which a single photographic type lens can be used to give magnifications up to about 40:1.

Photomacrography is closely related to photography using conventional photographic lenses. This is true particularly in the low magnification ranges. At magnifications greater than about 4:1, the modern large aperture photographic lenses do not perform well. Beyond this range, reverse mounting of the lens improves performance. Enlarging lenses, macro lenses, and small aperture photographic lenses (such as the Tessar design), when reverse mounted on the bellows, give excellent results up to the range of 15–20:1.

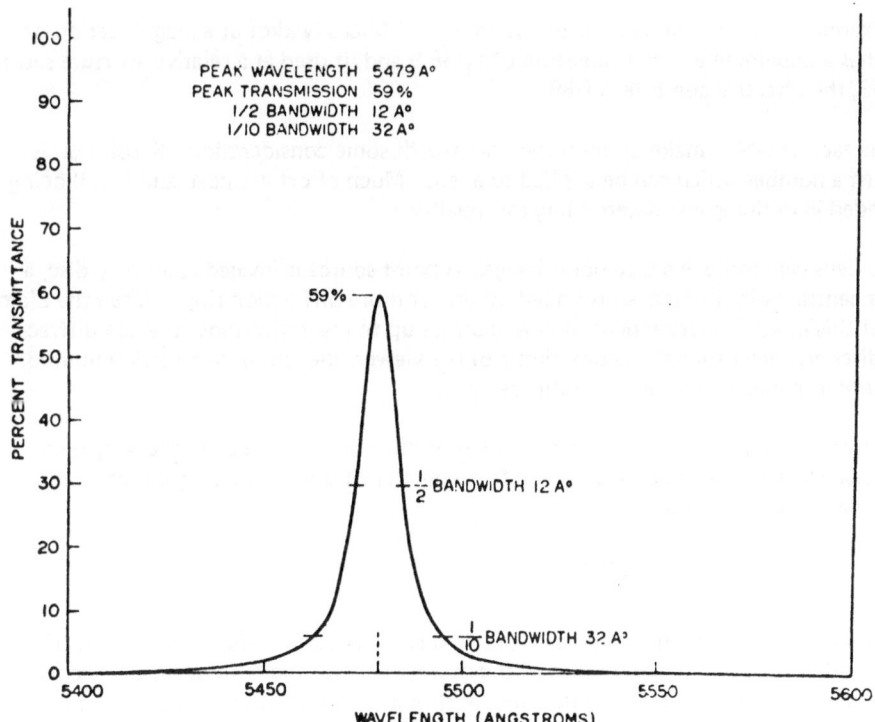

PEAK WAVELENGTH   5479 A°
PEAK TRANSMISSION 59%
1/2 BANDWIDTH   12 A°
1/10 BANDWIDTH   32 A°

Fig. 7. Interference Filter Spectrum (Baird Atomic)

At still higher magnifications, short focal length lenses from 16 mm motion picture cameras, again reverse mounted, give good coverage even of 4 x 5 plates up to 40:1 or slightly more.

One of the most convenient photomacrographic devices is the Polaroid MP4. The large camera format, long bellows extension, interchangeable lenses, and reflex ground glass focusing are ideal for this purpose. The stand and lights alone, when used with a 35 mm or medium format single lens reflex camera and bellows, provide a rigid framework for slides or other small negative films.

A major problem encountered in this type of photomacrography is the understanding of the aperture relationships which control exposure. Photographic lenses use the relative aperture system or f number which is defined as the ratio of focal length, F.L., to diameter of the aperture, d, by the following;

$$f = \frac{F.L.}{d}$$

when the lens is focused at or near infinity. However, the effective aperture, f_e, determines exposure when the lens is closer to the object than to the film plane. Effective aperture is a function of magnification and relative aperture according to the following;

$$f_e = (1+M)f$$

For example, a macro lens such as the 55 mm f/3.5 Micro Nikkor at a magnification of 10:1 has a maximum effective aperture of $f_e/38.5$ and, if used at a relative aperture setting of $f_e/8$, the effective aperture is f/88.

The factors which make up resolution are worth some consideration. Resolution is not just a number which can be applied to a lens. Much effort has been and is still being expended in defining and determining this quality.

No lens can produce a true point image. A point source is imaged as an Airy disc, a bright central point or area, surrounded by one or more diffraction rings. When the diameter of this image is independent of lens aperture up to magnifications at which diffraction discs are larger than the acuity limits of the viewer, the lens is "diffraction limited." Most microscopic objectives are in this category.

The old criteria for resolution depended upon the ability to separate the Airy discs from two closely spaced point sources. The diameter of the central bright spot of the disc is given by the following;

$$D = \frac{1.22\,\lambda}{n \sin \Theta}$$

where λ is the wave length of the light source, n is the index of refraction of the medium, and Θ is the half angle of acceptance of the objective lens. The term, n sin Θ, is called the Numerical Aperture, N.A., of the lens. Although it is not exact, in practice, it is found that the resolving power of a lens determined by this method is expressed reasonably well by the following;

$$R = \frac{\lambda}{2\,N.A.}$$

From this relationship the reason for using as short a wavelength of light and as large a numerical aperture as possible is obvious. It is also obvious why optical microscopy is limited in magnification by the physics of light rather than by limitation of equipment.

Some interesting arithmetic can be carried out concerning lenses and numerical aperture. A relationship between numerical aperture and the photographers effective aperture is given by;

$$f_e = \frac{M}{2\,N.A.}$$

In Table 1, some relationships of N.A., effective aperture, and resolving power are shown for several objective and conventional photographic lenses.

Another factor which is illustrated here is that the maximum resolving power of a lens is achieved at its maximum aperture. This is not the common experience with camera lenses where best resolution is attained usually in the center of the aperture range. There is often confusion between the depth of focus and resolution. Depth of focus is reduced by larger aperture while resolution is improved. The usual impression from photography is of greater sharpness from smaller apertures.

TABLE 1

NUMERICAL APERTURE RELATIONSHIPS

Objective Type	N.A.	Eqivalent Photographic Aperture	Resolving Power (um)	Depth of Focus (um)
55 mm F/3.5 Micro Nikkor at 10x	0.26	38.5	1.06	12.0
Achromat 10x	0.25	2.0	1.10	8.0
Apochromat 10x	0.30	1.7	0.92	5.5
Achromat 20x	0.50	1.0	0.55	2.0
Apochromat 20x	0.65	0.8	0.46	1.0
Achromat 45x	0.85	0.6	0.32	0.25
Apochromat 47.5x	0.95	0.5	0.29	0.10

Although it is not common practice, microscope objectives of the longer focal lengths, up to about 16 mm or 10X, can be used as photographic lenses for photomacrography. Curvature of field limits the area of highest resolution to a small central portion of the projected image. Yet, where maximum resolution of a small field is required or for use with small films such as 16 mm motion pictures, this offers some advantages. A highly corrected lens, operated at maximum aperture, will give a superior image. Any lens between the objective and the film can only degrade the image.

Normally, when short focal length objectives are used, an optical relay lens is required for convenient working distance. The real image formed by the primary lens becomes the object of a second lens. Chromatic corrections and field flattening functions can now be shared and higher apertures in the objective can be designed. With the additional lens, a compound microscope is formed [8,9,10, 11].

Magnification of the primary image does not result in an increase in resolution. It merely makes the resolution of the objective lens available to the microscopist's eyes. Because the magnification is not accompanied by increased resolution, there is a practical limit to the benefits of higher powered eyepieces. This has been variously and arbitrarily set between 500 and 1000 times the N.A. of the objective. The reason for the variability lies in the contrast of the image. A high contrast specimen will be improved up to nearly 1000 NA while a low contrast specimen may not be improved much beyond 500 N.A.

However, this discussion relates to photo-optics, and, as such, another of those physical limitation problems is found. The resolution limit in photography is determined by the size of the silver grain of the emulsion, or more properly, the size of the circle of confusion in the emulsion. Because the emulsion has a finite thickness, light may diffuse around the point source image, causing a diffusion halo. The size of the diffusion halo or circle of confusion is on the order of 0.015 mm. If M is the magnification on the film plane, then

$$\frac{M \lambda}{2 \text{ N.A.}} = 0.015$$

$$M \cong 60 \text{ N.A.}$$

which would indicate that the maximum photographic capability on the film for an air-spaced objective would be about 50X. However, this value is routinely exceeded. The reason lies in the complex nature of a microscopist's response to resolution.

The attempt here is to define the quality of sharpness that exists in a print. It differs for various purposes and within different areas of the same photograph. (Normally the sharpness decreases from the center toward the edge of a print). The quality varies with the observer too. A high contrast photograph will appear sharper than one of low contrast.

Considerable progress has been made in the past few years in developing new criteria for optical component performance evaluation [12, 13, 14] which can be correlated with image quality. Called modulation transfer function, the method considers all the components of the system, lens, film, enlarging or printing device, and the printing paper-developer combination. Ideally, each component is rated and the system performance is simply the product of the component ratings.

In order to determine a transfer function, a modulation pattern, consisting of a series of black and white bars of equal width and spacing, are transferred through the component being tested. The response is plotted as object/image brightness ratio vs. spatial frequency in lines/mm, Fig. 8. An 80% transfer would mean that, at that spacing of the bars, the brightness difference between light and dark was 80% of that of the specimen. As the lines per millimeter are increased, the response decreases approaching a limit at which the contrast between light and dark becomes so little that the individual bars are indistinguishable.

COLOR

Most work performed by metallographers is probably done in black and white. Possibly 80% or more of the photomicrographs taken by metallographers are of mounted, polished, and etched specimens. The critical problems are resolution of the structure and flatness of field. Because of the structure of lenses and the physical nature of light, some compromises are necessary for color. It is helpful to realize some of the problems, their origins, and current approaches to their solutions.

Color films do not have the flexibility of the human eye-brain compensating system. They require a light source which supplies a full spectrum of wavelengths in the proper ratios of intensity. Each type of color film is neutrally balanced to faithfully reproduce colors only when exposed under the specified lighting condition or color temperature.

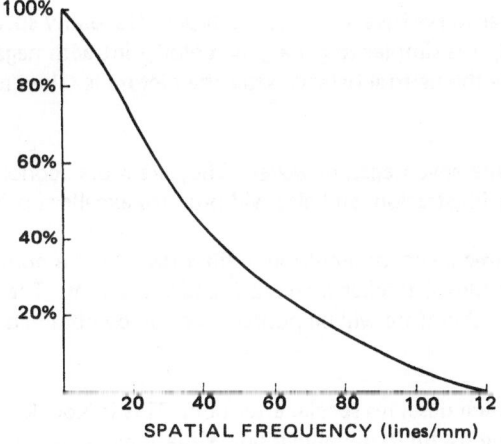

a) Spatial Frequency Response of a Lens

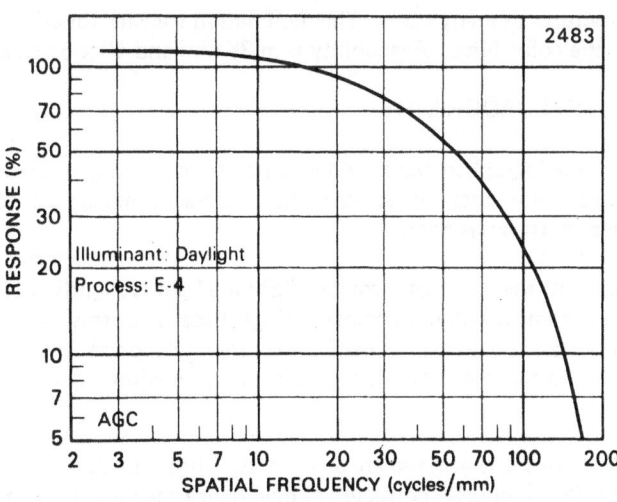

b) Spatial Frequency Response of a Film (Eastman Kodak Type 2483)

Fig. 8. Modulation Transfer Curves of a Lens and a Film

Furthermore, films are subject to sensitivity changes at different rates in the different color layers because of reciprocity failure, the non-linear response to extremes of time of exposure. Film manufacturers supply tables of times and correction filter recommendations to adjust exposures and color temperature for reciprocity effects.

There are two light balance types of color film, daylight and artificial. Ideally, with proper filtration, either can be used with any type of light source. Practically, it is better to use an artificial, or 3200°K, film with tungsten sources and an outdoor, or 5500°K, film with xenon illumination.

Color film also comes in positive and negative types. Generally speaking, if color reproduction is critical, it is simpler to get a good color print with negative films. More can be done to improve the neutral balance after the picture is taken than can be done with a positive slide.

Color slides have some advantages, however. They are more economical, can be used directly for teaching or illustration, and also will produce excellent prints.

Because there are three layers of emulsion, a filter layer, and a number of interleaving protective layers, color film is thicker than black and white film. The circle of confusion, and thus the resolution, therefore will be poorer than can be obtained with fine-grain black and white film.

There is one film which deserves special attention. This is Kodak 2483. It has certain properties which may be useful in metallographic situations. It is a positive, transparency film, balanced for 5500°K, slow speed, and critical in exposure. However, it has the finest grain structure of any color film and high contrast. Colors are exaggerated which helps to define structures in many cases. The modulation transfer function, Fig. 8, is the best of any of the color films. Availability is in 35 mm and 4" x 5" sheets.

OPTICS FOR PHOTOMICROGRAPHY

Generations of metallographers have produced fine photomicrographs routinely with achromatic objectives. Now there are apochromats, plan-achromats, fluorites, and plan-apochromats. There must be reasons.

In a simple glass lens, the index of refraction for blue light is slightly greater than for red light. A magnified image will have colored fringes because of this. Also, a simple lens does not produce an image in a flat plane. Ever more elaborate lens systems have evolved in effort to give the microscopist good resolution in white light and a flat image plane.

The most common objectives in use are achromats. These are chromatically corrected at two wavelengths but spherically corrected at one, hence the yellow green filters. Reasonably good color photomicrographs can be made if a UV absorbing filter is used. It is usually possible to interchange eyepieces and objectives of various manufacturers if the tube length is maintained.

Apochromatic objectives have been chromatically corrected at wavelengths in the blue, green, and red. Spherical correction is at two wavelengths. Performance in color photography is excellent because of these corrections. An added advantage is a slightly larger numerical aperture than for the equivalent magnification in an achromat and thus slightly better resolution capability.

Better chromatic correction comes with two limitations, however. First, the primary image of the objective has a high lateral color. Secondly, the primary image has high field curvature. Full chromatic correction is obtained only when these objectives are used with compensating eyepieces which have the same chromatic effect but in reverse order. The high field curvature can be partially corrected in the eyepiece. Thus, the field of highest resolution in the photographic image plane is smaller than that of an achromat of the same magnification.

The mineral fluorite, CaF_2, is used for internal elements of most apochromats. Because it is softer than glass it cannot be used on exposed surfaces. Also, lenses with fluorite elements are generally unsuitable for use with polarized light.

A series of objectives known as "Fluorites" have chromatic corrections between the achromatic and apochromatic. These are generally similar in performance to apochromats but less expensive.

More recently, a major advance in photomicrography has come about with the advent of flat-field objectives. More than simply objectives, these are systems. Each manufacturer introducing flat field optics has approached it differently. In all cases the improvement has involved thick lenses and compensating eyepieces. In some cases an additional lens has been added to the body tube above the objective for sharing corrections. This means that the objective — correction lens — eyepiece combination is critical and should not be disturbed by interchanging objectives or eyepieces.

For visual or photographic purposes, the improvement over older systems is spectacular. With these systems, the entire visible and photographic field can be maintained in sharp focus at the same time.

THE CAMERA

Finally, the image produced in the optical system must be relayed to a film.

An eyepiece will project a focused real image on a plane at a distance of 250 mm. The magnification of this image will be the product of objective and eyepiece magnifications. If a film was placed in this position, the whole microscope system would function as a simple, fixed-focus box camera.

Modification of the camera to permit focusing simpler, changing the magnification to get the field on a smaller film, adding a shutter, or inserting a photometer, is easily accomplished within this framework. The microscope camera has evolved much as the modern single-lens-reflex camera with built in exposure control has developed from the box camera.

It is important that nothing should be done to degrade the quality of the image which the microscope manufacturer has gone to such pains to produce. An example of one common error is found in the use of large format cameras supported on a separate stand. The bellows length is set and the camera moved over the microscope. Final focusing is done on the ground glass using the fine focus adjustment on the microscope. This changes the objective to object distance and in so doing, changes magnification and alters the optical correction characteristics slightly. At long bellows extensions and with large film formats, this is not generally noticed in the final print.

When short bellows extensions and small format cameras are used, the situation is different. In this case, refocusing with the fine adjustment would yield visibly increased image deterioration. To compensate for this, an auxiliary lens in the exit pupil of the eyepiece is used to refocus the virtual image at this point as a real image at the film plane. With an auxiliary lens of focal length precisely equal to the bellows extension length, placed at the eyepoint, no readjustment between visual and film planes of focus should be necessary.

Reporting of image magnification has caused much confusion. The factors are: objective magnification, eyepiece magnification, camera bellows length, and print enlargement factor. The final magnification is given by;

$$M_{total} = M_{objective} \times M_{eyepiece} \times \frac{bellows\ length}{250\ mm} \times M_{print}.$$

A simpler and more accurate method of determining the magnification is to photograph a stage micrometer using the same conditions and optics. At the time the final enlargement is made, a convenient unit of length is marked on the print or on the negative.

SUMMARY

There is a vast body of literature on light, optics, film, photomicrography, and all the other topics so briefly mentioned in this paper. As was stated in the beginning, optical microscopy has been developed to a state approaching theoretical limits. Yet, the practice is often considered somewhat of a mystery or at best an art. It need not be this way. The physics of light and of image formation are reasonable in application. It is hoped that this discussion can serve as the beginning of a more comfortable relationship with photo-optics.

REFERENCES

1. Roger P. Loveland *"Photomicrography, A Comprehensive Treatise"*, 2 Vols., J. Wiley & Sons, New York, NY, (1970).
2. R. Kingslake, *"Applied Optics and Optical Engineering"*, 5 Vols., Academic Press, New York, NY, (1965).
3. Eastman Kodak Co., *"Kodak Filters for Scientific and Technical Uses"*, Kodak Publication No. B-3, (1972).
4. Some sources are: Baird—Atomic, Inc. 125 Middlesex Turnpike, Bedford, MA, 01730; Karl Lambrecht Corp., 4204 N. Lincoln Ave., Chicago, IL 60618.
5. R. Westen, "Photomacrography with the Leitz Aristophot in Color and Monochrome", *The Microscope, 21,* 215—227, (Oct. 1973).
6. W.C. Hyzer, "Photography: Macro or Micro", *Research/Development, 26,* 22—25 (June 1975).
7. Eastman Kodak Co., "Photomacrography", Kodak Publication No. N—12B, (1969).
8. "Image-Forming and Illuminating Systems of the Microscope", E. Leitz, GmbH, Wetzlar, Germany.
9. H. Determan and F. Lepusch, "The Microscope and Its Application", E. Leitz, GmbH, Wetzlar, Germany.

10. Eastman Kodak Co., "Photography Through the Microscope", Kodak Publication No. P–2, (1970).
11. H.W. Zieler, "The Optical Performance of the Light Microscope", Microscope Publications, Ltd., London, England, Chicago, IL, (1972).
12. G.C. Higgins, R.L. Lamberts and R.N. Wolfe, "Validation of Sine Wave Analysis for Photographic Systems", Communication 2028, Kodak Research Laboratories, Fifth Conf. of Int. Comm. for Optics, Stockholm, 24–30, (Aug. 1959).
13. O.H. Schade, Sr., "Resolving Power Functions and Integrals of High-Definition Television and Photographic Cameras — A New Concept of Image Evaluation", *RCA Review,* 32, 567–609, (Dec. 1971).
14. Rodger P. Loveland, "Characteristics and Choice of Photographic Materials for Photomicrography", *The Microscope,* 19, 177–204 (Apr. 1971).

PHOTOGRAPHIC METHODS

L. E. SAMUELS *

INTRODUCTION

The analysis of the structure of a metal involves a number of discreet and often complicated and difficult steps, the last one of which in any examination of importance may be that of making a permanent record of the structure. This record is the means by which the results of the examination are communicated to others, so that the effectiveness of the whole operation is determined by how effectively this communication is carried out.

Early metallographers had to be content with making handdrawn sketches of the structures observed, and there are those who contend that this is still the best way of ensuring that students observe as well as see. Nevertheless, communications of this nature are not fully convincing, and metallography started to make real progress only when photographic methods of image recording became available. Sorby, who was undoubtedly the founder of metallography and a genius of the English era of the gentleman amateur scientist, soon devised methods of making photographs but only at low magnification using oblique illumination, photographs which these days would be described as macrographs. One of his last papers on metallography in which the structure of pearlite in steel was described for the first time still was supplemented by a sketch (Fig. 1(a)) because high magnifications were involved (2). Techniques soon improved, however, and the stage was soon reached where photography at high magnification and high resolution was possible (Fig. 1(b)) (3). Good standards had been achieved by the 1920's (4), and by the 1930's a degree of excellence had been achieved that can scarcely be bettered (5). The difference is that, in these early days, high standards were the preserve of the most skilled and dedicated. They are now open to all who care to try.

The reason is partly that the camera e.g., the photographic microscope has become much easier to use, but also is partly because photographic processes have been developed to the stage where they are reliable and straightforward. Nevertheless, it is still unfortunately true that much metallography, even when carefully carried out at the earlier stages, is ruined by poor photography. Sometimes this is due simply to careless work, but sometimes it is due to a lack of appreciation of the basic characteristics of the photographic processes. Consequently, the purpose

*Materials Research Laboratories, Melbourne, Australia, 3032.

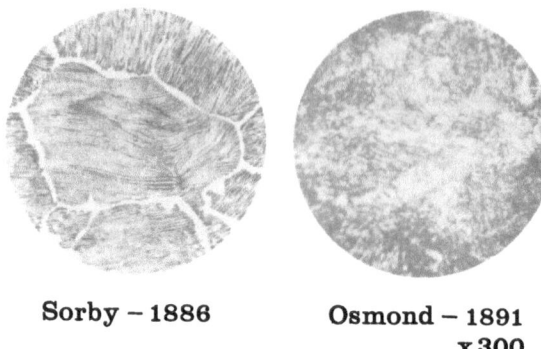

Sorby – 1886   Osmond – 1891
                        x 300

Figure 1. Early examples recording the structure of pearlite in steels. (a) A sketch
by Sorby (2) in a paper published in 1886 in which the structure of the "pearly'
constituent was described for the first time. (b) A photomicrograph published by
Osmond in 1891 (3), this being the first published photomicrograph resolving
pearlite. X800.

of the present paper will be to explore these basic characteristics, and to develop an
understanding of the strengths and limitations of the process as they apply to the
specific requirements of the recording of metallographic images. The recording of
optical images in black and white only will be considered. The recording of electronic images is a much simpler problem, but is one to which the same general principles apply, and the recording of images in colour will be considered elsewhere in
this symposium (6).

The idealized objective of scientific photography is to reproduce all of the tones
of the image in their original relationship. However, more than this usually is demanded consciously or unconsciously in metallographic practice. A metallographer
may want to compensate for defects in his optical system, and this usually means
compensating for flare which degrades contrast. He may wish to compensate for
defects in specimen preparation, which usually means adjusting for less-than-optimum
etching or perhaps hiding preparation artifacts such as polishing scratches. Finally,
a sensitive and sensible metallographer will want to produce an aesthetically pleasing
result, partly perhaps because it pleases him but primarily in the hope that it will
convince the viewer of the thoroughness and reliability of the metallographer's work,
so reinforcing the communication of the information that he has taken so much
trouble to create. Preparing an aesthetically pleasing result may take some skill but
the rest takes only care and understanding.

NEGATIVE – POSITIVE PROCESSES: NEGATIVES

The classical photographic process involves two stages, the first involving the
preparation of a transparent negative and the second the preparation of an opaque
positive print from this negative. Negative quality is the basis of final print quality,
since there is a definite limit to what can be done to redeem a bad negative; it is
certainly not possible to introduce into the print information that is not contained
in the negatives. The preparation of a negative also introduces the greatest technical

problems. Three major variables are involved, namely, choice of negative material (colloquially known as *negative emulsion**),choice of exposure, and choice of development procedures.

Most subjects can be described in terms of a range of tones of black and white. With some, the range of tones may be short and with others it may be long; with some, the tones may be uniformly distributed between the extremes and with others they may be compressed at one end or the other. In the simplest terms, the requirement is to convert the tone scale of the subject into an opposing tone scale in the negative.

Unfortunately many problems arise to complicate this simple picture, namely:
(a) The tone scale of the negative is inherently fixed whereas that of different object images may vary widely.
(b) The subject may have a tone scale longer than that which the negative material can handle satisfactorily.
(c) The range of tones produced in the negative may be greater than that which the positive print will accept.
(d) Negative emulsions vary in colour sensitivity and hence in the manner in which a range of colour tone is rendered as a range of tones of black and white.

It is perhaps some consolation that metallographic subjects are comparatively simple ones from these points of view. They are often black and white, or have such a small range of colour tones that they can readily be represented by a black-and-white tone scale. Further, proper operation of the microscope optics frequently requires that a limited range of colours be used in the system. Finally, the range of tones is usually short enough to be handled readily by available negative emulsions . There are, admittedly, some subjects with reasonably wide tone ranges best exemplified perhaps by specimens which contain extensive areas of a non-metallic phase of comparatively low reflectivity. Cast irons containing graphite are one example; specimens containing a surface scale or oxide layer represent an even more difficult example. Both will be used as illustrations later.

Basic Characteristics of Negative Emulsions

The fundamental factors underlying the handling of negative emulsions is still best described by the so-called *characteristic curve* described by the photographic pioneers Hurter and Driffield many years ago. The blackness of the processed photographic image is measured by determining the ratio of transmitted to incident light, I_t/I_o; this ratio is called the *transmittance* and its reciprocal the *opacity*. The logarithm of opacity is then called the *density* D. If density is plotted against the logarithm of total exposure E (the total light energy that was incident on the negative, i.e., the product of incident light intensity I and the exposure time t), a curve of the type shown in Fig. 2 is obtained for a particular set of development conditions.

---

* The active layer of a negative is not truly an emulsion, but a dispersion of small crystals in a transparent medium.

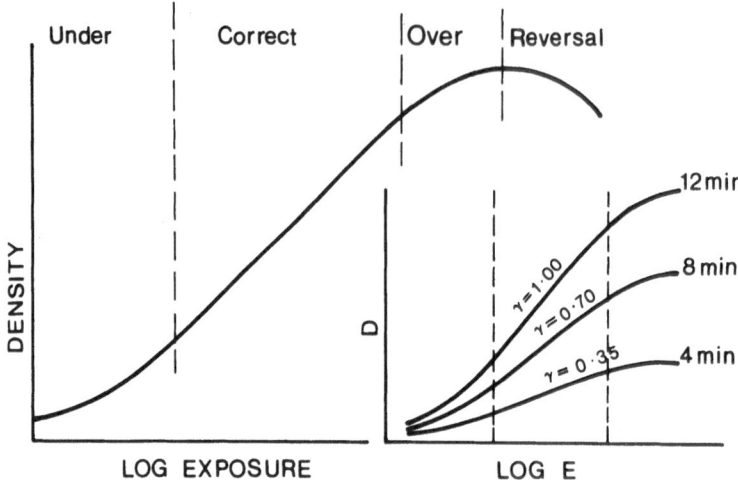

Figure 2. Diagrammatic illustration of the characteristic curve of a photographic emulsion, with insert showing the effect of development time.

This curve may be divided into four parts. They are, starting from the left of Fig. 2:

(i) A concave toe, for which each doubling of exposure results in an unequal increase in density and for which the separation of the (dark) tones consequently is poor. This is a region of *under exposure*.

(ii) A straight line portion for which each doubling of exposure results in an equal increase in density and for which the separation of tones is a maximum. This is the region for which the tones of the image are reproduced in proportion to their original relationships, and is the region of *correct exposure*.

(iii) A convex shoulder for which each doubling of exposure results in an unequal increase in density and for which the separation of the (light) tones is poor. This is the region of *over exposure*.

(iv) Beyond the shoulder there is a drooping region where an increase in exposure results in a decrease in density. This is known as *reversal* or *solarization,* but should not be encountered in metallography.

Note that the curve does not intersect the zero point of the scales, that is, some negative density is produced even with no exposure. This is known as *fog.* Moreover, there is a threshold exposure required to produce a density above fog.

The curve shown in Fig. 2 is somewhat idealized in that emulsions in practice do not always show such a clearly defined straight-line section. It is good enough for the present purposes, however, to assume that they do. Moreover, in general photography and with modern negative materials it is possible, and even desirable, to work in the toe region of the characteristic curve. However, in metallography it is easily possible to work on the straight line portion of the curve and it is best generally to attempt to do so. Variations from this principle might well be left to those highly skilled in photographic techniques. The subsequent discussion will be based on this premise.

In this event, the important feature of the characteristic curve is the slope of the straight-line portion. The tangent of the angle of slope is known as the *contrast* or *gamma* (γ) value of the emulsion. The subject tones will be strictly proportional reproduced in their original relationship when $\gamma = 1$ (that is, subject contrast and negative contrast will be equal), and in a more contrasty relationship when $\gamma > 1$ (negative contrast will be greater than subject contrast). The latter is desirable in metallography and a gamma of about 1.3 seem to be optimum. γ values of 2.0 or more can be attained but gives prints which are excessively harsh for general work; even gammas above 1.5–1.6 give results which generally appear too harsh.

Other important features of a negative emulsion that are described by the characteristic curve are its *speed* (a measure of the value of E needed to produce a negative density at the start of the straight-line portion) and *latitude* (the range of exposures which lie on the straight-line portion). Figures are available from the manufacturers for all of these factors which describe quantitatively the characteristics of an emulsion. However, these figures may be prepared with outdoor photography in sunlight in mind and then do not necessarily apply quantitatively to the conditions applying in metallography.

Each characteristic curve illustrates the behaviour of the material under one set of conditions, and a different curve will result if any change is made in the light source, developer, time of development, and temperature of development. Generally, the shape of the curve does not alter so much as its position. It is possible in metallography easily to keep the light source and type of developer constant, or at least to standardize on a very limited range of light sources and developers, which leaves developing time and developer temperature as the variables requiring attention.

When identically exposed emulsions are subjected to different development times, a series of different but related characteristic curves is obtained, such as those illustrated in the bottom right-hand corner of Fig. 2. All curves have the same general shape described earlier; even more important, the same exposure values remain in the same basic sections of the curve. That is, under exposure is not lifted into the straight-line section by longer development, and shorter development does not bring over exposure down to the straight-line section. *Exposure is the only factor which determines where a negative ends up on the characteristic curve.*

However, development time does affect the slope of the straight-line section, that is, the contrast. Gamma increases, generally at a declining rate, with increase in development time (Fig. 3(a)). Likewise, the temperature of the developing solution affects contrast only, but has a significant effect (Fig. 3(b)); development time has to be decreased when the temperature rises to ensure the same contrast. Moreover, serious physical damage to the emulsion may occur with higher developer temperatures, and temperatures about 25°C should not be used without special precautions.

Finally, it is assumed in the above discussion that decrease in the intensity of the incident light can be compensated for linearly by increasing the exposure time. This is true for any negative emulsion over a certain range of light intensities, but eventually breaks down; more than linear increase in exposure time is then required to

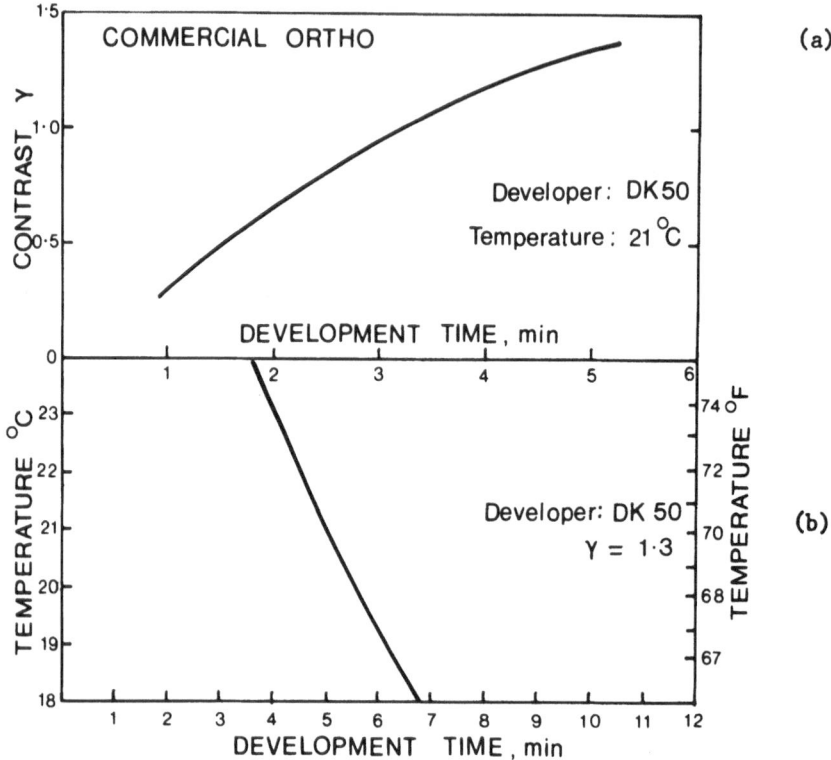

Fig. 3. Effect of development time (a), and developer temperature (b) on the contrast developed in a negative.

obtain the same negative density. This is known as *reciprocity failure*. This would seem to be a possibility in microscopy because the intensity of the light incident on the negative tends to be low, but no difficulty is encountered in practice with modern equipment and black-and-white negative emulsions. It is a problem however, in colour photomicrography.

Practice — Choice of Negative Emulsion

Metallographic examinations carried out at magnifications below about X1000 under bright-field illumination invariably use achromatic objective, and adequate performance of an achromatic objective is obtained only if the wavelength of the illuminating light is kept within a comparatively narrow range. Traditionally this range has been selected to be in the region of 5000—6500A°, which is a yellow-green light to which the human eye is most sensitive. The first requirement of a negative material therefore is that its spectral sensitivity should match this range. This is found in the type of emulsion known as *orthochromatic.*

A wider range of wavelengths could be used with fluorite and apochromatic objectives commonly used at magnifications of X1000 and above, but there is no point to this except in the most special circumstances. Practice is simplified if these objectives are also operated in the same yellow-green light and orthochromatic negative emulsions used again. This emulsion has the following additional advantages:

(i) It is fine grained, and hence can be used to record fine detail and stand considerable enlargement

(ii) Gammas in the desired range can be developed in a grade known as *commercial ortho*

(iii) It has sufficient latitude easily to accommodate the range of tones found in most metallographic specimens and still allow for some errors in exposure.

The *speed* of orthochromatic emulsions is comparatively slow, but this is so for all fine-grained emulsion. Moreover, reasonably long exposures (certainly up to several minutes) are acceptable in metallography.

This emulsion is also optimum for many special methods of illumination which produce an image in the normal yellow-green range. It is not suitable for techniques which use the full range of the visible spectrum, the commonest example being polarized light. *Panchromatic* emulsions must then be used; this will not be discussed specifically, but the same principles apply as for orthochromatic emulsions.

There was once a tradition that glass-backed negatives, known as *plates,* had to be used for photomicroscopy rather than plastic-backed negatives, known as *films,* apparently because it was thought that films were not flat enough to record a sharp image over their full area. If so, this is completely erroneous because the image produced by a microscope is sharp over a considerable distance, many times the possible irregularities of a reasonably well supported film. Plates suffer from serious disadvantages in their bulk for storage and potential for breakage.

Practice — Choice of Exposure

The point has already been made that correct exposure of the negative is one of the most important factors in photomicrography. The exposure needs to be chosen so that all of the important tones of the subject, and preferably all of the tones, are exposed on the straight-line portion of the characteristic curve. An incorrectly exposed negative cannot be redeemed by any trick of development or printing.

This can be a difficult matter to judge in general photography because the intensity of light incident on the subject is highly variable, the reflectivities of various parts of the subject also differ, and the range of subject tones is wide. But it is a much simpler problem in metallography, provided that the microscope is used at a series of standardized magnification as is recommended by the ASTM Standard Method (7), and other conditions such as the nature of the illuminating sources, the filters used in the system and the optical components in use are kept constant at each magnification. This is not only easy to do but is plain good practice; the optimum combination should be selected for each standard magnification and then adhered to. Under these circumstances, the intensity of light incident on the specimen is highly constant and, since the reflectivity of specimens of a particular alloy group vary very little, the intensity of light at the image plane is sufficiently predictable to permit the development of a system of standardized exposures (8).

The principle of a standardization system can be illustrated by considering the case of a normalized plain-carbon steel photographed in bright-field illumination, the microstructure consisting of light-toned ferrite in a matrix of dark-toned unresolved pearlite (Fig. 4(c)). A photomicrograph of this type of subject is always prin-ted so that the ferrite appears just off-white. The appearance of the pearlite areas consequently is automatically determined by the contrast of the original specimen and that of the photographic processes, provided that the ferrite was exposed suf-ficiently far enough up the straight-line portion of the characteristic curve for the pearlite areas also to be exposed on this portion. This same exposure would then be equally acceptable for specimens containing different proportions of ferrite and pearlite, including those containing no ferrite and those comprised entirely of fer-rite. The validity of this argument is confirmed by the series of photomicrographs illustrated in Figs. 4 (a)—4(c). The same exposure would also be equally acceptable for any steel which contained constituents whose reflectivity allowed them also to be exposed in the straight-line section of the characteristic curve (Fig. 4(d)). The same type of argument can be developed for any other alloy system, but the expo-sure required to expose the brightest constituent to the required point on the straight-line portion of the characteristic curve would be different because of the different reflectivities of the metals concerned.

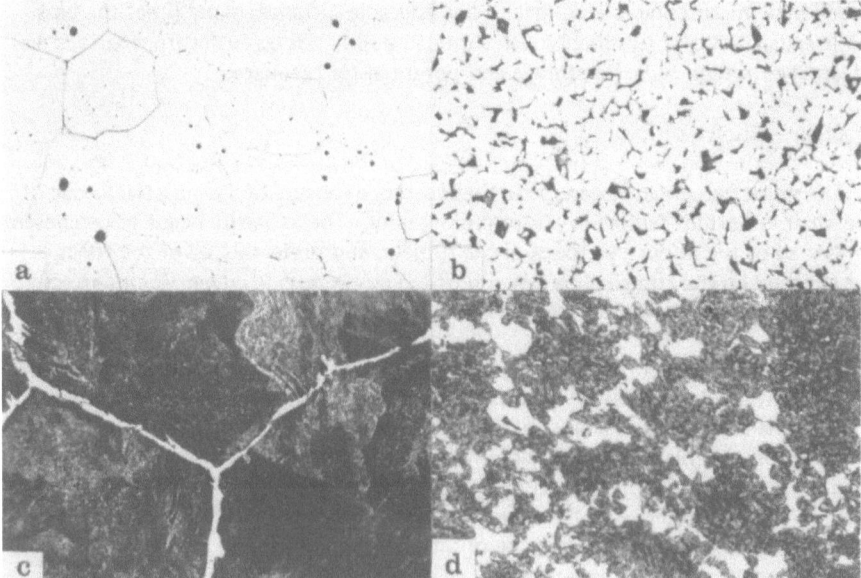

Fig. 4. Series illustrating the validity of the concept of standardized exposures. All specimens were photographed with the same exposure, the negatives were developed together and printed under precisely the same conditions. X250. (a) ingot steel, (b) mild steel, (c) normalized medium — carbon steel, (d) ferrite and tempered mar-tensite. All are optimum prints in spite of the considerable differences between the specimens.

This concept is merely an application of the standard photographic technique of determining exposure from the high-light intensity of the image.

A simplified method of standardizing the exposures of a microscope based on this concept is as follows: Choose a specimen that encompasses the widest range of tones likely to be encountered in the work of the particular laboratory. For example, the grey cast iron shown in Fig. 5 would be suitable for many laboratories handling a wide range of specimens since the tone range ferrite-pearlite-graphite is about the widest likely to be encountered. A specimen with a smaller tone range might be acceptable to others. Set up the microscope at a suitable magnification with optimized optical conditions and expose a set of negatives (or step negatives*) and develop them under controlled conditions aimed at producing the chosen gamma value. Prepare the best possible print from each negative on a medium grade of printing paper, as in Fig. 5. On the basis of these prints, select the negative which gives acceptable tones in all three phases with minimum printing time. This will be the negatives for which the ferrite has been exposed just enough to ensure that the pearlite and graphite have also been exposed on the straight-line portion of the characteristic curve. The density of the ferrite areas of this negative can then be adopted as a reference high-light density.

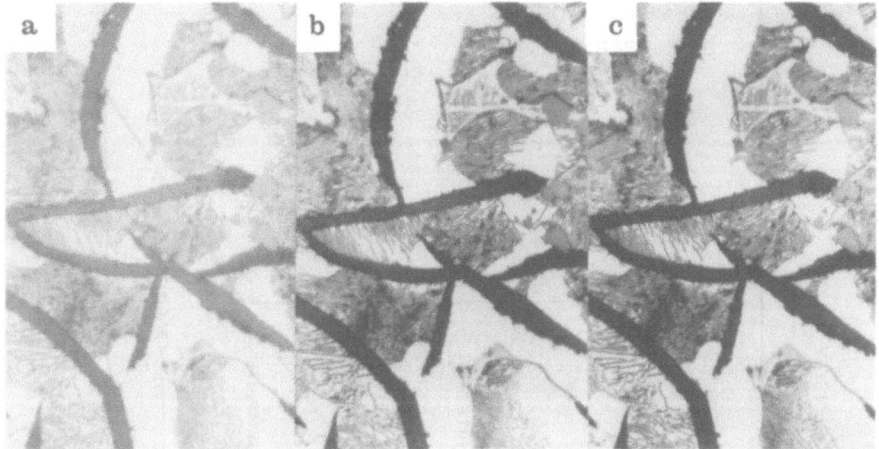

Fig. 5. Grey Cast Iron. X500. Series illustrating a method of determining an appropriate high-light negative density. Exposure times were progressively increased, the negatives developed to the chosen gamma value, and printed on the one grade of paper to the best possible result. (a) Negative exposed 5.2 sec. Under exposed — poor tonal representation of matrix constituents. (b) Negative exposed 20 sec. Good exposure — good tonal representation of all constituents. (c) Negative exposed 48 sec. Still a good exposure, but an excessively long printing time was required.

---

* Step negatives usually are prepared by removing the dark slide of the negative holder in steps, exposing at each step. This has the serious disadvantage that incremental exposures are not equivalent to a single exposure of the same total duration (the *intermittency effect*). A more satisfactory method is to cut a slot in a dark slide; strips of the negative can then be exposed individually by withdrawing the dark slide in steps.

These principles can be illustrated in a more quantitative way if a densitometer is available. Portion of the characteristic curve for the chosen emulsion and developer determined by exposure in the microscope is shown in Fig. 6. The densities of the ferrite and graphite areas of the negative for Fig. 5 (a) are indicated on this figure, and it is apparent that they have been exposed on the toe of the characteristic curve. The density figures for the negative of Fig. 5 (b), on the other hand, indicate that both have just been exposed on the straight-line portion of the characteristic curve. The negative for Fig. 5(c) has also been exposed on the straight-line portion, but further up this portion; the negative is unnecessarily dense. The ferrite density of the negative for Fig. 5(b) is this taken as the reference high-light density (D = 1.6). Details of this quantitative approach are given in references (8) and (9).

A specimen of aluminum (unetched) is then prepared as representing the highest specimen reflectivity likely to be encountered, and another set of negatives is prepared at each magnification and optical arrangement which it is intended to use. The standard exposure for a particular optical arrangement is taken to be that which produces a negative density the same as the high-light density (ferrite areas) of the reference micrograph described in the preceding paragraph. This comparison can be made by eye but is better made with a densitometer, again if one is available.

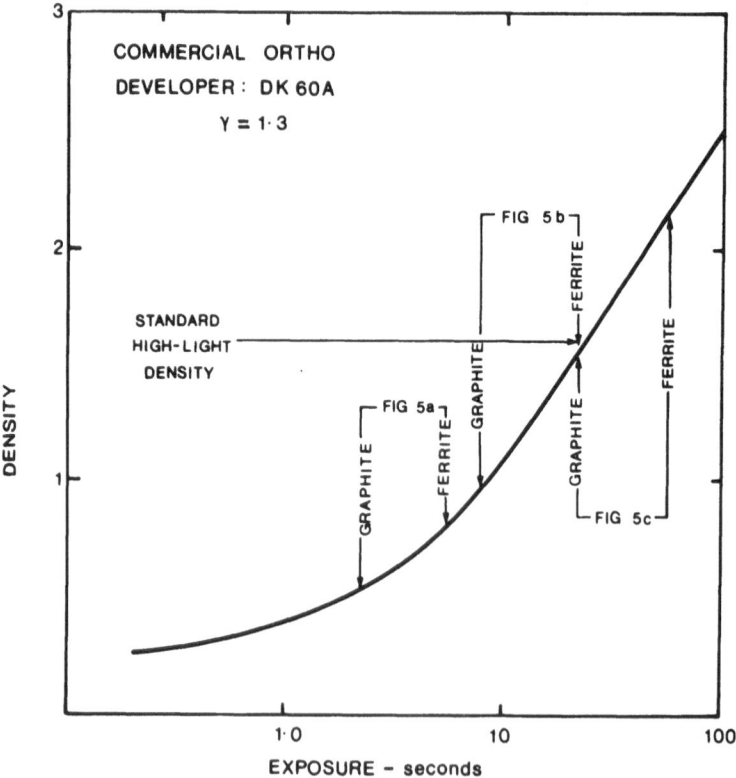

Fig. 6. Negative characteristic curve illustrating the exposure conditions used for Fig. 5.

The procedure can be repeated for unetched specimens of all alloys likely to be handled and an exposure factor determined for each group; this needs to be done at only two or three microscope arrangements since the exposure factor will be the same over the full operating range of the microscope. Since a precision of no better than 10% is required, most alloys will fit into a few exposure groupings (Table 1). It is then possible to develop a calibration and exposure table for a microscope, portion of such a table being set out in Table 2. The setting up of this table is actually a simple procedure, being no more than a systematized version of classical methods of carrying out trial exposures. Once done it is done for once and all.

TABLE 1

Exposure Factors For Alloy Groups

Metal or Alloy	Exposure Factor
Al; Ag; Zn; Cd; Brass (10–45% Zn)	1.0
Sn; Sb	1.2
Cu; Ni; Brass (0–10% Zn)	1.3
Bronzes (Sn or Al); Steel	1.5

TABLE 2

Calibration and Exposure Table

Microscope Serial No.: TR9869
Illuminating Source: Strip Filament
Light Filter: Interference—550
Negative: Commercial Ortho

Magnification	Objective	Ocular	Aperture Stop[1]	Bellows Extension	Exposure, Secs 1.0[2]	1.2	1.3	1.5
100	12.5 x Achro.	5 x Hyg.	3.0	210	4.5	5.5	6.0	7.0

[1] For complete filling of back element of objective
[2] Alloy group exposure factor

Additional exposure correction factors may also have to be determined, the most important being one to compensate for variations in aperture stop setting. Microscope aperture stops are usually fitted with an arbitrary angular scale. The marking that ensures that the back element of a particular objective is just fully filled with light can be determined and recorded, and thereafter any change in aperture can be considered as a percentage reduction of this figure (Table 3). Modern microscope objectives are best normally operated at, or close to, full aperture so that the correction factors in Table 3 are really only a reminder that some adjustment in exposure is necessary if the aperture is reduced.

TABLE 3

Exposure Correction for Aperture Stop Setting

Reduction in Aperture Stop, %	Increase in Exposure %
5	10
10	25
15	40
20	60

It may be necessary also to determine calibration factors for alternative light sources and other illuminating conditions. Modern microscope light sources are, incidentally, remarkably uniform in output intensity throughout their life. It may also be necessary to determine exposure factors for special conditions; for example, some etching methods deposit a film of low reflectivity on the specimen surface and a special calibration factor would then be required. Such factors have of course only to be determined once, but even so the great bulk of metallographic practice is covered by the procedure outlined in Table 2.

Photoelectric meters are available for determining exposure and some modern microscopes have meters of this type built into them, some even being arranged automatically to control exposure. Apart from any consideration of whether the cost of these devices is justified, which is extremely doubtful at best in metallography, they can be downright misleading when they are based on a measurement of the average intensity of a considerable area of the image. For example, the average intensities of the four specimens illustrated in Fig. 4 differed by a factor of nearly two and a device of the type under consideration would have recommended considerably different exposures. Yet the optimum exposure in all four cases is the same. Meters of this type will give an acceptable result only where the tonal range of the specimen is much less than the latitude range of the negative emulsion in use. Exposure meters which measure a spot high-light intensity are sound in principle.

Exposure meters may however be useful for monitoring the constancy of the illumination system of a microscope. The meter reading for a standard specimen, such as as-polished aluminum, could be recorded for a selected optical arrangement and checked from time to time whenever doubts arose. Moreover, they are applicable to circumstances when the reflectivity of the specimen surface or the absorption in the optical system is highly variable, such as in polarized light. A scheme of standardized exposures cannot then be applied, but selection of exposure is still better based on high-light intensity than average intensity.

It must be conceded, however, that the situation is quite different in transmitted light microscopy where the light absorption by the specimens is highly variable. Some means of measuring the actual intensity of light at the image plane is then very useful.

Practice — Development

Manufacturers of negative emulsions recommend a number of developers suitable for each of their emulsions. All are highly reliable, and one should be selected which will develop the chosen negative emulsion to the contrast gamma desired. The soundest advice that can be given is that, once a developer has been chosen, the manufacturer's recommended conditions of use should be adhered to rigidly. Many years of experience has shown that, if exposure has been brought under control by some method such as the system of standardized exposures just described, unsatisfactory negatives are invariably due to failure to follow this advice in the darkroom.

As was illustrated in Fig. 3, the first thing that must be considered is the relationship between developer temperature and development time since both must be controlled to produce the required contrast. Either the developer must be kept at a suitable constant temperature and a standard development time appropriate to the required gamma chosen, or the developer temperature must be measured and the development time adjusted by the use of a curve such as Fig. 3(b). The darkroom operator must never be allowed to guess when the negative has been developed.

To achieve the desired results in practice, it is better to develop in tanks than in dishes, and preferably the developer tank should be held in a constant temperature bath. Remember also that excessively high developer temperatures may cause damage to the emulsion, and that transfer from a hot developer to a comparatively cold wash may also cause physical damage. All processing solutions must also be kept in good condition, and the manufacturer's advice must be followed here as well.

Manufacturers also recommend a particular agitation regime during development. This must also be adhered to, not only to ensure full reproducible development but also to ensure even development. Operators don't like following an agitation regime; it is tedious, but it has to be done.

Attention to detail at the finishing stages is also necessary if serious negative defects are to be avoided. Immersion in an acid stop bath is desirable both to stop the development process promptly and to prevent carry over of (alkaline) developer into the (acid) fixer bath. Thorough fixing and washing to the manufacturer's recommendations are obviously essential to ensure proper storage life. Practical problems arise particularly during washing.

Firstly, a free flow of mains water usually must be used for washing, and suspended matter in this water readily attaches itself to the soft emulsion of the negative; the negative will be spoilt if this material is allowed to dry into its surface, and the delicate surface of the negative may be damaged if an attempt is made to wipe it off. It is imperative to filter the wash water when there is any risk of this problem; a 20 μm filter is adequate. Secondly, drops of water adhering to the negative surface may easily dry off to produce so-called *water marks.* The drying drop causes the emulsion to swell around its edge and to shrink at its center, producing a crater-like defect which prints as a light ring with a dark center; no amount of rewashing will remove a water mark once it has formed. They can be avoided by immersing the negative in dilute solution of a wetting agent after washing and by then immediately sponging off both surfaces before drying.

NEGATIVE – POSITIVE PROCESSES: PRINTING

Basic Characteristics of Printing Papers

Two basic types of printing paper are available (*chloride* and *bromide*) and both are equally satisfactory for printing metallographic negatives. Bromide papers are faster but have to be handled in more subdued light; they are also grainier, although this is generally of no practical consequence in metallography. They are necessary for enlargement and often are preferred for contact printing. A number of types of paper surface are also available, but only a glossy finish normally would be considered for contact printing to ensure maximum discrimination of fine detail in the print. Enlargements intended for display purposes are another matter, but will not be considered here.

The emulsion of a printing paper is basically similar to that of a negative, but the spectral sensitivity is different to permit handling under brighter lighting conditions. The performance of the emulsion can be described by a characteristic curve similar to that shown in Fig. 2 for negatives. In general, papers have higher contrast gammas than negatives but smaller latitude. However, contrast cannot be altered by control of development, because the development of papers must be taken to finality to obtain full rich blacks. Instead, a range of grades of paper with varying contrast gammas is made available by manufacturers.

The most appropriate ways to characterize a printing paper are therefore by, firstly, its gamma value and, secondly, the range of negative densities that it will print properly, that is, by the difference in density between the darkest and lightest areas of a negative that will print on the straight-line portion of the characteristic curve of the paper. The latter is known as the *exposure scale* of the paper. Representative values of gamma and exposure scale for the grades of paper normally available are listed in Table 4.

TABLE 4

General Characteristics of Grades of Printing Paper

GRADE	DESCRIPTION	CONTRAST (γ)	EXPOSURE SCALE*
0	Extra Soft	1.3	> 1.4
1	Soft	1.5	$1.2 - 1.4$
2	Normal	1.7	$1.0 - 1.2$
3	Contrast	2.0	$0.8 - 1.0$
4	Extra Contrast	2.8	$0.6 - 0.8$
5		3.3	> 0.6

* Expressed as the difference in negative densities that can be printed on the straight-line portion of the characteristic curve of the paper.

It is expected in general photography that grade 2 paper will properly print a well exposed and developed negative, but grade 3 paper seems to be more suitable for the higher contrast results aimed for in metallography, with grades 1, 2 and 4 being used for adjustments. The primary problem is that the tonal range recorded in the negative may be much longer than that which can be recorded in a print. This point is best illustrated by a metallographic example.

The subject illustrated in Fig. 7 is a tin bronze with a surface layer of corrosion product and some shrinkage cavities; the negative densities of the different areas of the negative are listed in the caption. When this negative is printed on grade 3 paper so that the bronze areas have a just perceptible deposit of silver (appears just off-white), the corrosion layer prints as a full black. So also do the specimen mount and the internal cavities. Thus the important corrosion layer is not distinguished in the print (Fig. 7(a)). The printing time may be reduced so that the oxide and mount are exposed within the grey scale of the paper and thus be discriminated in the print (Fig. 7(b)), but then the bronze is not printed at all so that no detail is developed in this area and the evidence that the corrosion has followed the dendritic segregation pattern is lost. On the other hand, grade 1 paper has a wide enough exposure scale for this negative, and the three primary areas of the subject are properly discriminated (Fig. 7(c)), although the bronze area appears rather muddier than aesthetically would be desirable. A sacrifice thus has to be made to convey the information of greatest importance.

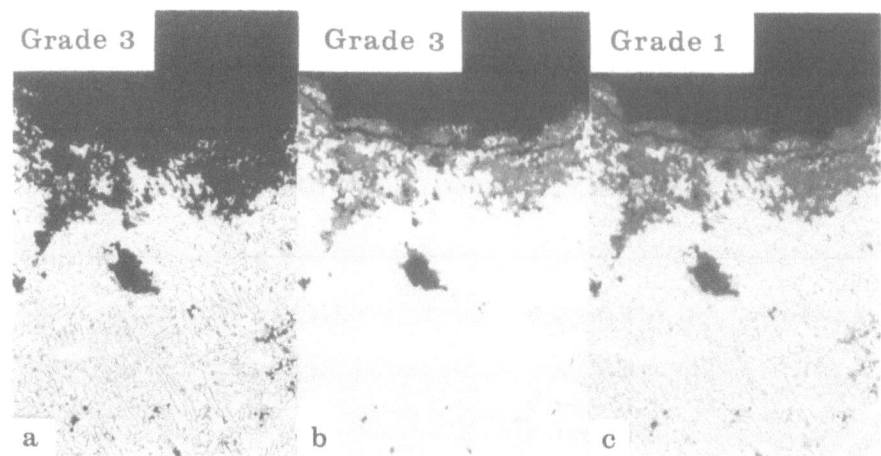

Fig. 7. Tin Bronze with Layer of Corrosion Product. X100. Series illustrating effect of grade of printing paper for a specimen with a large tonal range. Negative characteristics:

	Negative Density	Corrected Negative Density Scale
Bronze Matrix	1.8	
Corrosion Product	0.9	1.1
Mount	0.6	1.3

(a) Printed on grade 3 paper, (b) Printed on grade 3 paper, but with shorter exposure time than (a), (c) Printed on grade 1 paper.

These considerations can be quantified by determining the negative density scale of the features of interest i.e., the difference in density of the relevant areas of the negative and adding to this figure the value 0.2, a correction found necessary by experience. The appropriate printing paper is the one whose exposure scale covers this corrected negative density scale. Thus, in the Fig. 7 example, the corrected negative density scales for bronze to corrosion product and bronze to shrinkage cavity are about 1.1 and 1.3, respectively, and Table 4 indicates that grade 1 paper is the most contrasty grade whose exposure scale will cover these figures.

Another type of compromise that might be necessary, this time in the reverse direction, is exemplified in Fig. 8, which illustrates a cast iron with flake graphite and a matrix of ferrite, pearlite and phosphide eutectic. The corrected negative density scale for this negative indicates that grade 1 paper would be necessary to print detail in the graphite flakes as well as the matrix (cf. figures in caption to Fig. 4 with those in Table 4), and experiment shows that this is indeed so (Fig. 8). However, the resultant print is muddy and rather unattractive (Fig. 8(a)) but would have to be used if conveying information about the internal structure of the graphite flakes was important. If it were not, however, detail in the flakes might be sacrificed by using a grade 2 paper, which covers the corrected negative density range of the

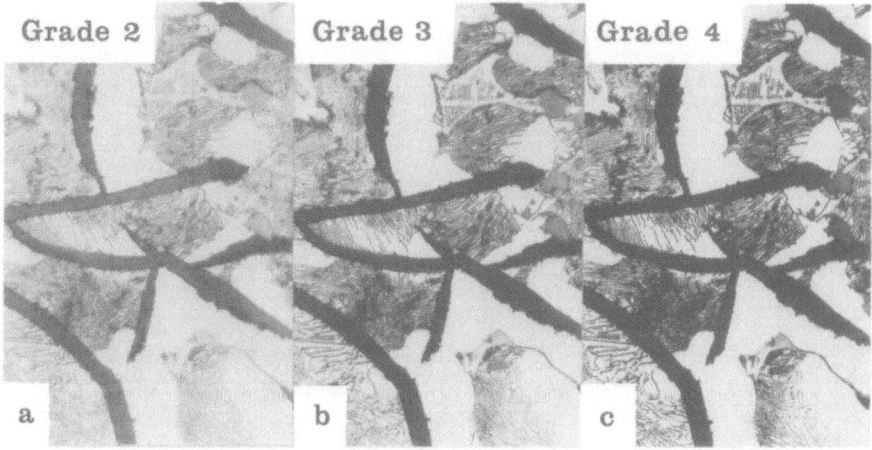

Fig. 8. Grey Cast Iron. X500. Series illustrating a compromise in choice of grade of printing paper for a specimen with a large tonal range. Negative characteristics:

	Negative Density	Corrected Negative Density Scale
Ferrite	1.7	
Phosphide eutectic	1.3	0.6
Pearlite	1.1	0.8
Graphite	0.9	1.0

(a) Printed on grade 2 paper, (b) Printed on grade 3 paper, (c) Printed on grade 4 paper.

matrix phases and so prints them with a good crisp tonal range, producing a more acceptable overall print (Fig. 8(b)). A more contrasty paper again, however, loses detail in both the graphite and the matrix (Fig. 8(c)). Thus there is no one correct print in this case, but rather an optimized print matched to the information that most needs to be conveyed, and the tastes of the metallographer.

A subject with a more normal tone range for metallographic subjects is illustrated in Fig. 9. The corrected negative density range in this case is so small that a fully proportion tonal range can be obtained in a print on grade 4 paper (cf. figures in caption with those in Table 4). But these considerations indicate only the most contrasty grade of paper that may be used to obtain a fully proportional print. Many would think that the print on grade 4 paper was excessively harsh (Fig. 9(c)), a feature that is more apparent in a larger area of print than that shown in Fig. 9. A softer grade of paper can reasonably be used if desired and many would prefer a print on grade 3 (Fig. 9(b)), although most would agree that one on grade 2 (Fig. 9(a)) was too soft. Possibly the negative contrast was too high for this subject; even if so, it is still generally best to standardize negative development and adjust contrast during print.

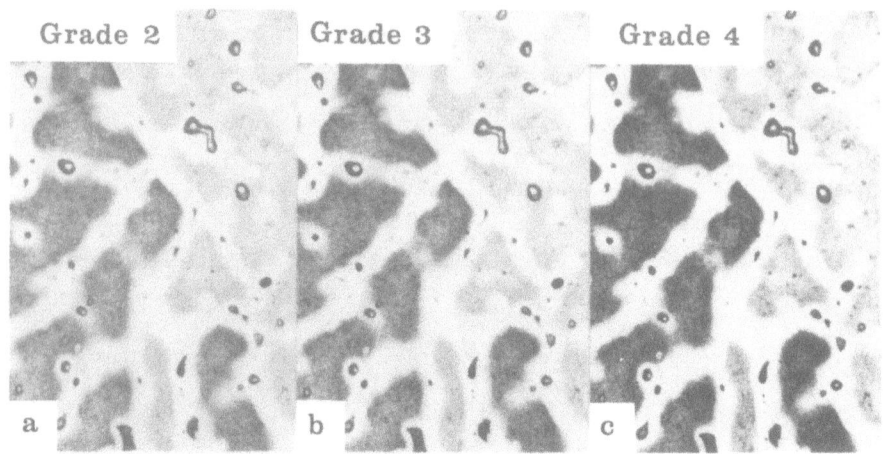

Fig. 9. Cast Leaded Brass. X500. Series illustrating choice of grade of printing paper for a specimen with a small tonal range. Negative characteristics:

	Negative Density	Corrected Density Density Range
Interdendritic Regions	1.4	
Lighter grain	1.1	0.5
Darker grain	1.0	0.6

(a) Printed on grade 2 paper, (b) Printed on grade 3 paper, (c) Printed on grade 4 paper.

Practice — Choice of Paper Grade

It has just been shown that the grade of paper needed to print proportionately the density range of a negative can be determined in a quantitative way with a densitometer. Although this is a convenient way by which to illustrate the principles involved, it is much simpler in practice to choose a grade of paper by experience and trial, bearing in mind the principles just illustrated.

Practice — Printing Time

There is no hope of arranging matters to achieve closely standardized printing times; too many variables are involved and the latitude of printing papers is too narrow. Optimum printing time must therefore be determined by trial and error in each case. However, it is possible to keep printing times within a reasonably narrow range when the procedures for standardizing negative exposure and development recommended earlier are adopted. The sensitivity of printing materials to small variations in exposure time also means that it is desirable to use a reliable timing device to control exposure when numbers of similar prints of the one negative have to be produced.

Practice — Development

Again, manufacturers recommend developing procedures for their products and these must be adhered to for best results. In particular, the full recommended development time must be used otherwise the black tones of the print will be impaired. The practice of operators terminating development when they think that the print looks right should be severely discouraged.

Until comparatively recently, papers universally were developed in trays, and then processed through a stop bath, fixer and wash. They cannot then be inspected in a bright light to assess quality until fixing is well advanced, and this involved a cycle time of at least 3 min., a lengthy time if many trial exposures are required. More recently, a simple and more rapid *stabilization process* has become available.

The special papers required have characteristics similar to normal papers, but a developing agent is incorporated in the emulsion during manufacture. The exposed paper is processed through a machine where it is treated with an alkaline solution to complete the developer, immediately followed by a stabilizer solution. The print is then nearly dried by squeeze rollers and emerges fully developed in about 10 sec. This print will keep for some years, provided that it is kept away from heat and humidity, but a standard fixing and washing treatment is necessary for archival keeping qualities.

TRANSFER REVERSAL PROCESSES

A revolutional photographic recording process known formally as the *diffusion transfer reversal process* (DTR), but more commonly by the trade name "Polaroid", has been available for some time and has found wide acceptance in many areas of general photography. Although it is widely used in certain areas of scientific photography, particularly in the recording of electronic images, it has found only limited application in high-quality metallography. Its usefulness in metallography, and the limitations to its usefulness, will now be explored.

Advantages

The DTR process has a number of obvious advantages which at first sight make it appear to be extremely attractive. It is unnecessary to invest in expensive photographic darkrooms; it is unnecessary to make up and maintain a series of photographic solutions; a messy and fairly complicated developing process is replaced by a simple clean one and, as a consequence, the operating staff require virtually no training; a final print is produced within a few seconds instead of several hours at the very least. On the face of it, it is surprising that the conventional negative-and-print process has not been completely supplanted. However, the process also has some severe limitations and it is these limitations that must now be examined so that a metallographer can balance them against the attractions that have just been mentioned.

Disadvantages

Firstly, a rather limited range of DTR materials is available in any one form so that some of the considerable flexibility of the conventional photographic process is lost. Nevertheless, the range of film sizes available happens to be adequate for metallographic purposes, and all metallurgical microscopes can easily be fitted with suitable holders for one or other of the available types.

The range of available emulsion types which could be considered for metallography is listed in Table 5, which includes comparative information for a conventional orthochromatic negative material that is commonly used in metallography. Note that most DTR emulsions are panchromatic in spectral sensitivity and so can be used with most light source and filter combinations. They are also mostly much faster than conventional orthochromatic negative materials. This is an advantage if the microscope is subjected to vibration.

TABLE 5

DTR Emulsion Types Suitable for Metallography

Type Reference	Available As	Spectral Sensitivity	Speed As a Number	Granularity	Contrast
42	Roll	Panchromatic	200	Medium	Normal
47	"	"	3000	"	"
107	Pack	Panchromatic	3000	Medium	Normal
52	Sheet	Panchromatic	400	Medium	Normal
57	"	"	3000	"	"
55P/N	"	"	50	Very Fine	"
51	"	Blue Only (4600 A)	200	Medium	High
Commercial Ortho Negative	Sheet Film	Orthochromatic	20	Fine	High

The next group of disadvantages result directly from the very fact that the DTR system is a reversal process. It follows firstly that there is little latitude in exposure time. This problem is compounded by the high cost of the material, so that the process could become a very expensive one indeed if the correct exposure was determined as a routine by trial and error methods. Some exposure estimation or standardization system consequently needs to be established; exposure shutters need to be reliable and accurate; and the microscope light source must either be stable or be monitored. All of these precautions can be fairly readily achieved, by methods similar to those discussed earlier, but the effort has to be made. About one good print in three or four exposures is necessary to place the DTR process on an equal economic footing with conventional processes, all things including labor being considered. It also follows that the contrast obtained in a DTR print is not adjustable, so that compensation cannot be made for variations in the specimen and specimen preparation.

Next, a positive reversal process produces but a single print. This may be quite adequate if the photomicrograph is being taken merely for record purposes, but often multiple copies are required for publications and sometimes an enlargement is required. However, one of the available DTR materials produces a positive print and a negative transparency simultaneously (emulsion Type 55 P/N listed in Table 5). The negative emulsion is an extremely fine grained one and, after a simple fixing and washing process, lasts indefinitely and can thereafter be printed like any other photographic negative.

The disadvantages discussed so far are really minor ones, and would be of no great consequence if it were not for the major failing of the DTR process from the present point of view, namely, its comparatively poor intrinsic contrast. Indeed the contrast is so poor as to place the process out of all consideration for high-quality metallography.

COMPARISON WITH THE CONTRAST OF CONVENTIONAL PROCESSES

The accompanying photomicrographs give a guide to the relative magnitude of the contrast obtainable by DTR and conventional processes. The specimens illustrated are a pearlitic steel and are of about average contrast as metallographic subjects go. A particular field is shown in Fig. 10(a) after photographing with a fluorite objective using a conventional commercial orthochromatic negative material. The same field is shown in Figs. 10 (b) — 10(d) after photographing under the same conditions using a range of DTR emulsions. In all cases the prints represent the best results that can be obtained. Figs. 10(b) and 10(d) are actually photographic reproductions of the DTR print, but closely represent the original.

Fig. 10(b) is a result representative of standard DTR emulsions; it was actually taken with Type 52 emulsion but almost identical results would be obtained with Types 42, 47, 57 and 107 emulsions, all of which are classified in Table 5 as having medium contrast. The results obtained with the Type 51 emulsion, which is classified as having high contrast, is shown in Fig. 10(d). The contrast now more closely approaches the conventional results, but unfortunately one of the characteristics

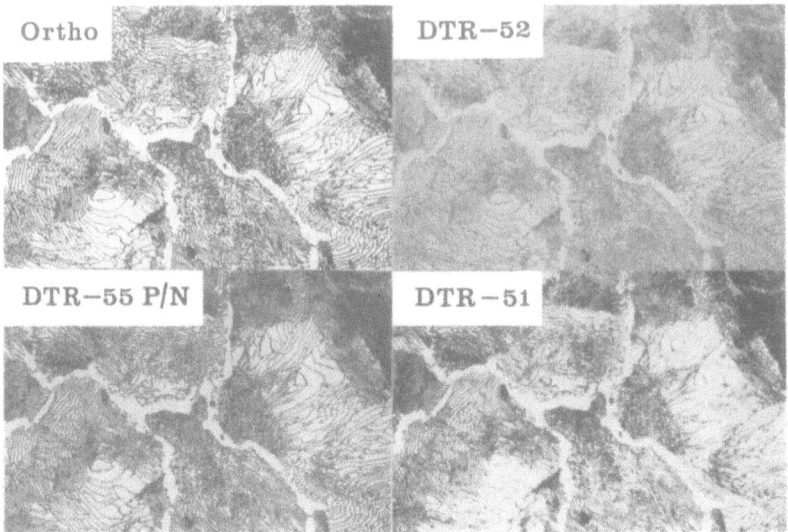

FLUORITE OBJECTIVE

Fig. 10. Comparison of contrast obtained by DTR emulsions with a fluorite objective. (a) Photographed by conventional processes, (b) DTR Type 52, (c) DTR Type 55 P/N; negative printed on conventional high-contrast paper, (d) DTR Type 51, using blue filter. Microscope Objective: 80X, 0.95NA fluorite. Magnification: X1000.

of this emulsion is that it is sensitive only to very short wavelength (blue) light; even if light containing longer wavelengths is used in the optical train only the appropriate short wavelength components actually contribute to image formation. This imposes a severe limitation in metallographic practice, for the following reasons.

Most objectives supplied with commercial metallographs are achromats, this being almost invariably so for dry objectives intended to be used at magnifications of X1000 or less. Very poor results are obtained when the high-contrast Type 51 emulsion is used to record the image produced by even high-quality achromatic objectives (Fig. 11) because they are just not designed to produce a sharp image in blue light. The results obtained with most fluorite-type objectives are acceptable, fluorites being in effect achromats with improved chromatic performance; this type of objective was used in preparing Fig. 10(d). Good results really are obtained only with apochromatic objectives, when a record quite comparable to that of conventional photographic processes is obtained (Fig. 12). It is necessary to point out again, however, that apochromats are normally supplied only as high-aperture objectives intended for use at high magnifications, and usually as oil-immersion objectives. The net result is that, in practice, it is possible to use the Type 51 emulsion effectively under only a limited range of metallographic conditions, the range varying according to the nature of the series of objectives with which a microscope is equipped.

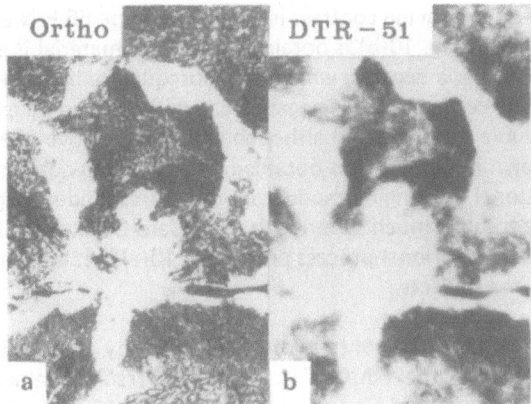

ACHROMATIC OBJECTIVE

Fig. 11. Comparison of contrast obtained by DTR Type 51 emulsion with an achromatic objective. (a) Photographed by conventional processes, (b) DTR Type 51, using blue filter. Microscope Objective: 80X, 0.85NA achromat. Magnification: X1000.

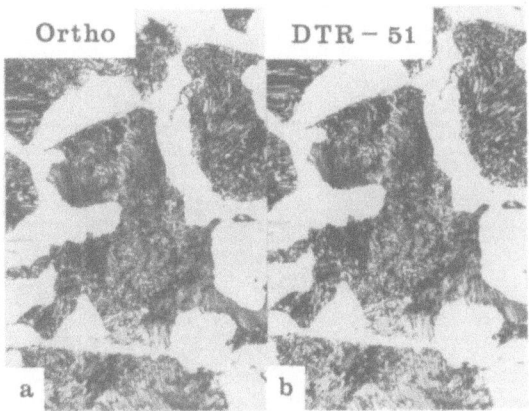

APOCHROMATIC OBJECTIVE

Fig. 12. Comparison of contrast obtained by DTR Type 51 emulsions with an apochromatic objective. (a) Photographed by conventional processes, (b) DTR Type 51, using blue filter. Microscope Objective: 58X, 0.95NA apochromat. Magnification: X1000.

Another, but more minor difficulty, arises from the short exposure latitude of the Type 51 emulsion. It follows that exposure control is even more critical than usual for DTR materials and that it may not be possible to expose correctly all regions of a subject having a wide range of tone values; the latter effect is partly evident in Fig. 10(d).

A possible way of improving contrast is to use the Type 55 P/N emulsion. A positive print similar to Fig. 10(b) is obtained when this material is exposed at an ASA 50 rating, as well as a negative which can subsequently be printed by conventional methods. The contrast of this negative can be improved by exposing at ASA 25 rating (i.e., double the exposure) although an initial positive of even lower contrast than that shown in Fig. 10(b) is obtained. A maximum-contrast print then obtained by conventional printing methods from the corresponding negative is shown in Fig. 10(c). The result is much superior to a standard DTR print (Fig. 10(b)) but is still inferior to a conventional-process print (Fig. 10(a)); it may however be quite acceptable with some subjects.

It is possible to improve the contrast of a DTR print by copying it by conventional photographic methods. For example, Fig. 13 shows a result obtained when the low-contrast DTR Type 52 print of Fig. 10(b) was copied in this way, the final print quality being similar to that obtained by the conventional photomicrographic process. However, a reasonable degree of photographic skill is required to upgrade the contrast of a DTR print in this way, and it is only likely to be successful with subjects with good initial contrast. The process could not be considered as a routine but it might be considered, for example, when a field of importance has been lost and only a DTR print is available.

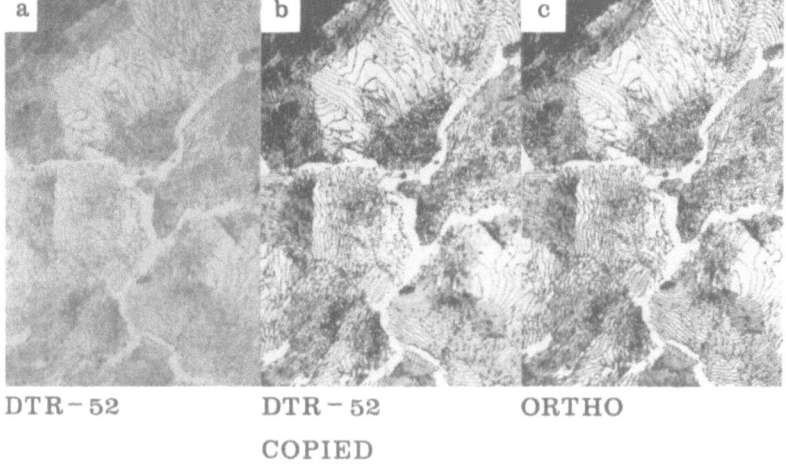

Fig. 13. The DTR Type 52 print illustrated in Fig. 10(b) copied by convention photography to improve contrast.

Field of Application of the DTR Process in Metallography

Thus, although there are a number of reasons why it is tempting to use the DTR process in a metallographic laboratory, whether it is possible actually to do so or not depends largely on the following factors:

       1. The inherent contrast of the specimens that are handled. The process may be acceptable if most of the specimens have high intrinsic contrast but not if most have medium or low contrast.

2. The types of objective with which the microscopes available are equipped. It may be possible to use the high-contrast Type 51 emulsion extensively if a wide range of fluorite or apochromatic objectives is available but not if most of the objectives available are achromats.
3. The standard of contrast that is acceptable. The process may well be quite acceptable if the need is merely to keep a permanent record of microscopical observations.
4. The frequency with which multiple prints are required. The process is much less attractive if multiple copies have to be prepared with any regularity.
5. The equipment available. The DTR process will at least produce a permanent record when a photographic darkroom is not available.

CONCLUSIONS

It was stated at the outset that the idealized objective of scientific photography was to reproduce all the tones of the image in their original relationship. The subsequent discussion strayed from this ideal, although not far from the concept that the subject tones should be represented proportionately.

The practical objectives of metallographic photography really must be based on the overriding concept that a photomicrograph is intended to convey certain information, and this it must do above all else with clarity and impact. Aesthetic considerations definitely are secondary, but still are important because they add to the impact and to the confidence of the viewer that all the important information has been discovered and truely and reliably conveyed.

The key to good conventional photomicrography is to record all the information of importance proportionately in the negative image. Once this has been done, much flexibility is possible in selecting the information conveyed in the final print and the manner in which it is conveyed. In turn, the key to a good negative is correct exposure, and this is a comparatively simple matter to ensure in metallography.

The photographic materials now available are highly reliable and well designed to meet the needs of metallography. The characteristics of these materials are well understood, and the preparation of good photomicrographs is simply a matter of applying conscientiously procedures compatible with these characteristics.

ACKNOWLEDGMENTS

The author gratefully acknowledges much helpful advice and discussion with Mr. C.P. Johnson, Head of the Scientific Photography Section, Materials Research Laboratories.

REFERENCES

1. H.C. Sorby, *J. Iron Steel Inst.*, 31, 255, (1887).
2. H.C. Sorby, *J. Iron Steel Inst.*, 28, 140 (1886).
3. F. Osmond, *J. Iron Steel Inst.*, (i), 160 (1891).

4. R.H. Greaves and H. Wrighton, *"Practical Microscopical Metallography"*, 1st edition Chapman and Hall: 1924 (London).
5. J.R. Villella, *"Metallographic Techniques for Steel"* American Society for Metals: 1938 (Cleveland).
6. R.J. Gray, this publication.
7. *"Standard Methods of Preparation of Micrographs of Metals and Alloys"* ASTM Designation: E2–62.
8. L.E. Samuels, T.O. Mulhearn and R.M. Robb, *Metallurgia,* 57 (1958).
9. R.C. Gifkins, *"Optical Microscopy of Metals"*, Elsevier: 1970 (New York).

APPLICATIONS OF COLOR IN METALLOGRAPHY AND PHOTOGRAPHY*

R. S. CROUSE[†], R. J. GRAY[†], and B. C. LESLIE[†]

INTRODUCTION

In the visual experience of most humans color is commonplace and one naturally finds color photographs more realistic than black and white. However, this is generally not true in the field of metallography where the images one normally encounters are black and white or shades of gray. Therefore, the occurrence of color in a microstructure may tend to cause the metallographer to accept it on the basis of aesthetic appeal rather than reality.

Yakowitz[1] has proposed two general use categories for color in metallography that enables evaluation of the utility of a colored microstructure. They are (1) contrast enhancement which utilizes the eye's natural ability to distinguish color hues and (2) color to indicate chemical composition. It is these categories that must be addressed to justify the somewhat increased expense of carrying color metallography to its logical conclusion, the color photographic image.

The availability of high quality microscopes, cameras and relatively inexpensive commercial photographic processing bring color metallography into the reach of even the smallest of laboratories. Processing of positive transparencies for slides is reliably done by numerous film manufacturers with quite acceptable results.

MICROSTRUCTURAL COLOR

Some of the techniques used to produce color in microstructures are described in this paper and are presented in Table 1. In each of the four techniques, a film or overlay produces, or is directly related to, the formation of the color. The production of the color is the result of film characteristics such as thickness, morphology, index of refraction, microtopography, and composition. Additional roles are played by the characteristics of the substrate and the impinging light, i.e., polarized bright field, dark field, sensitive tint. In almost every case, the observed color is a synergistic result.

This paper presents only a select few examples to illustrate the listed techniques. Where applicable, reagents and conditions are listed in the appendix.

---

* Research sponsored by Union Carbide Corporation under contract with Energy Research and Development Administration.
† Oak Ridge National Laboratory, Oak Ridge, Tennessee 37830 USA.

Figure 1

Plate A. Ceramic to cermet braze joint tinted by heating on a laboratory hot plate.

Plate B. Yttrium as polished and chemically stained (see appendix). The specimen was photographed on ' type B'' film using bright-field illumination, carbon arc light source.

Plate C. These two photomicrographs of $UC_{1.6}$ illustrate the combination of two modes of color production on a single specimen. The left one is chemically stained and reveals the primary U_2C_3. The right one is the same field in polarized light which shows the $UC-UC_2$ eutectic matrix and the grain structure of the U_2C_3. Photographed with carbon arc illumination on "type B" film.

Plate D. Zirconium containing hydride anodically stained using Picklesimer's method.

Plate E. Sigma phase in type 310 stainless steel. Photographed on "type B" film, bright-field illumination, carbon arc light source.

Plate F. Electrolytically etched aluminum (see appendix) displayed in normal (left) and rotated (right) sensitive tint. Carbon arc light source, "type B" film.

Plate G. Mid-plane cross section of a pyrolytic carbon coated nuclear fuel particle, 300 μm diameter. Sensitive-tint illumination was used to accentuate the graphite (yellow). The photographic film had a daylight emulsion and the light source was a Xenon lamp on a shielded, remotely operated metallograph.

Plate H. Reactor grade "needle-coke"graphite. This was photographed on a research metallograph in the as-polished condition using rotated-sensitive-tint illumination. The photographic film had a daylight emulsion and the light source was a Xenon lamp.

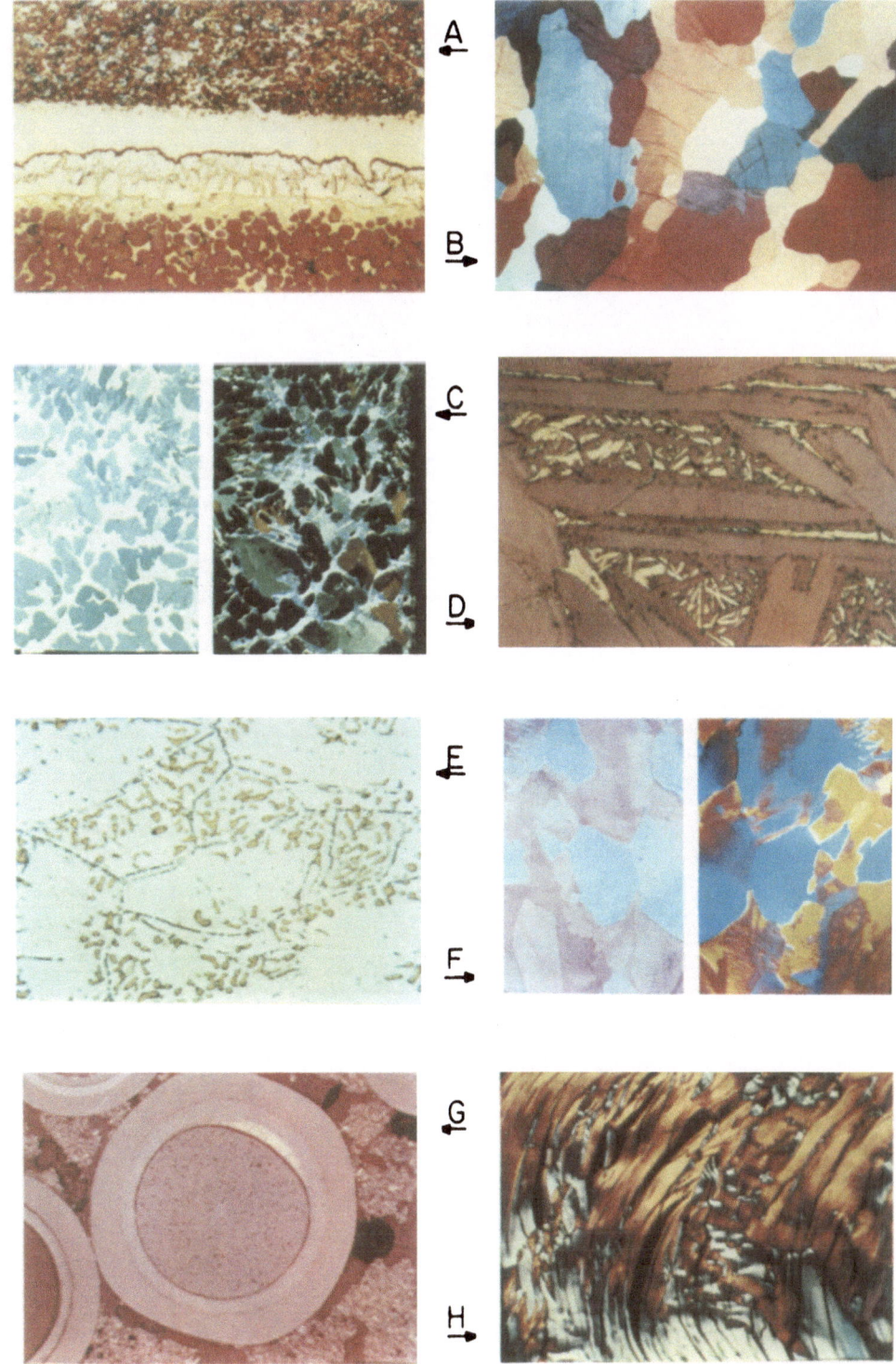

TABLE I

METALLOGRAPHIC COLOR

Technique	Examples
Heat tinting or Selective Oxidation	Carbon steel, brazing alloys cermets, oxides of Ta, Nb, Cu, etc.
Chemical	UC_x, Yttrium, sigma phase
Electrolytic	Al, Nb, Ti, Zr
Film deposition	Most multi-phase materials

Heat Tinting

The most elementary method of applying an interference film is the single expedient of heat tinting. This is usually done by heating a polished specimen in air in a furnace or perhaps, simply on a hot plate. Obviously, little control can be exercised, but occasionally spectacular results are obtained. Combinations of dissimilar materials, such as ceramics and cermets, sometimes respond favorably to heat tinting, an example of which is shown in Fig. 1(a).

Selective oxidation used in fundamental studies of selected metal surface reactions to a controlled oxygen atmosphere at elevated temperatures has been reported [2-5].

These reported oxidation patterns on single crystals have been used by Gray et al [6] to demonstrate similar metallographic phenomena. Numerous pictorial examples are given in this work, one of which is repeated here in Fig. 2. Black and white reproduction fails to do justice to the spectacular color but merely indicates the patterns which are directly related to rates of oxidation.

Chemical Coloring

The production of color by chemical means is accomplished by dipping or swabbing a specimen just as one might attempt a grain boundary etch. The interference film formed may be an oxide or some complex reaction product involving components of the etching solution and the various grains and/or phases of the specimen. The film is likely to be somewhat delicate and easily damaged so care must be taken to preserve it during washing and drying.

Two examples of chemical coloring for contrast enhancement are found with molybdenum and yttrium, yttrium being shown in Fig. 1, color plate B. The film is formed by dipping in the proper reagent (see appendix) and being careful not to touch the surface. The grain structure is revealed in color in bright field illumination.

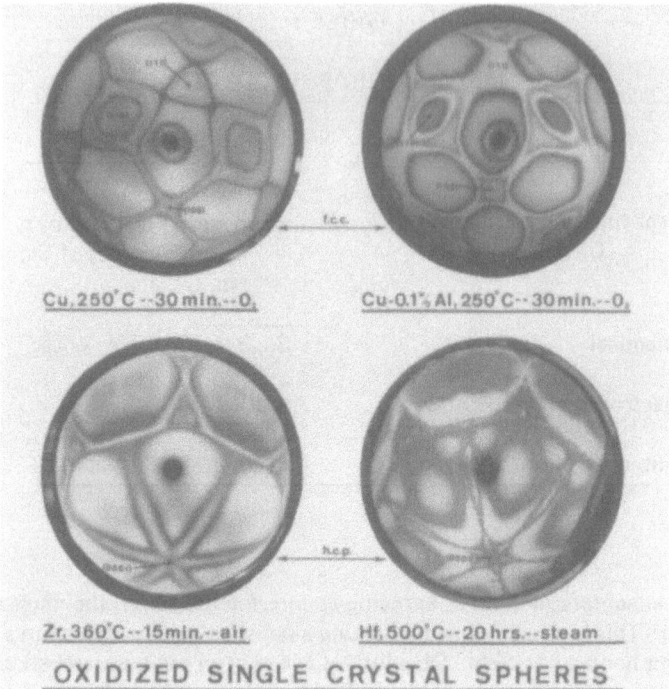

Cu, 250°C --30 min.-- O_2 Cu-0.1% Al, 250°C --30 min.-- O_2

Zr, 360°C -- 15min.-- air   Hf, 500°C -- 20 hrs.-- steam

OXIDIZED SINGLE CRYSTAL SPHERES

Fig. 2. Patterns developed on copper and copper alloy single crystal spheres for basic studies of oxidation mechanisms.

Chemical analysis, x-ray diffraction, and metallography often combine to present a clear understanding of a microstructure. An example of this is found in the uranium-carbon system which has been extensively reported [7-9]. An alloy of uranium and carbon of just under stoichometric UC_2 composition, ($UC_{1.6}$), was found by x-ray diffraction to contain UC, U_2C_3, and UC_2. By chemical staining the sesquicarbide (U_2C_3) is revealed and by viewing in polarized light the matrix is further seen to be UC and UC_2. This is illustrated in Fig. 1, color plate C (see appendix).

In the past ten years some very significant work has been done by Beraha [10-16] in selective coloration of phases in steels and stainless alloys. His techniques fall in the "indication of chemical composition" category. He used the electromotive properties of the alloys and phases to cause films of sulfide, oxide-phosphates, molybdates, etc., to selectively precipitate on either the second phase or matrix phase. By carefully selecting reagents, he can color specimens almost at will. This is very important in the metallography of steels.

Electrolytic Anodization

The production of interference films by electrolytic means provides a way of controlling the films precisely so that particular colors may be reproduced time and again. In his work with zirconium alloys, Picklesimer [17] established electrolytic anodization

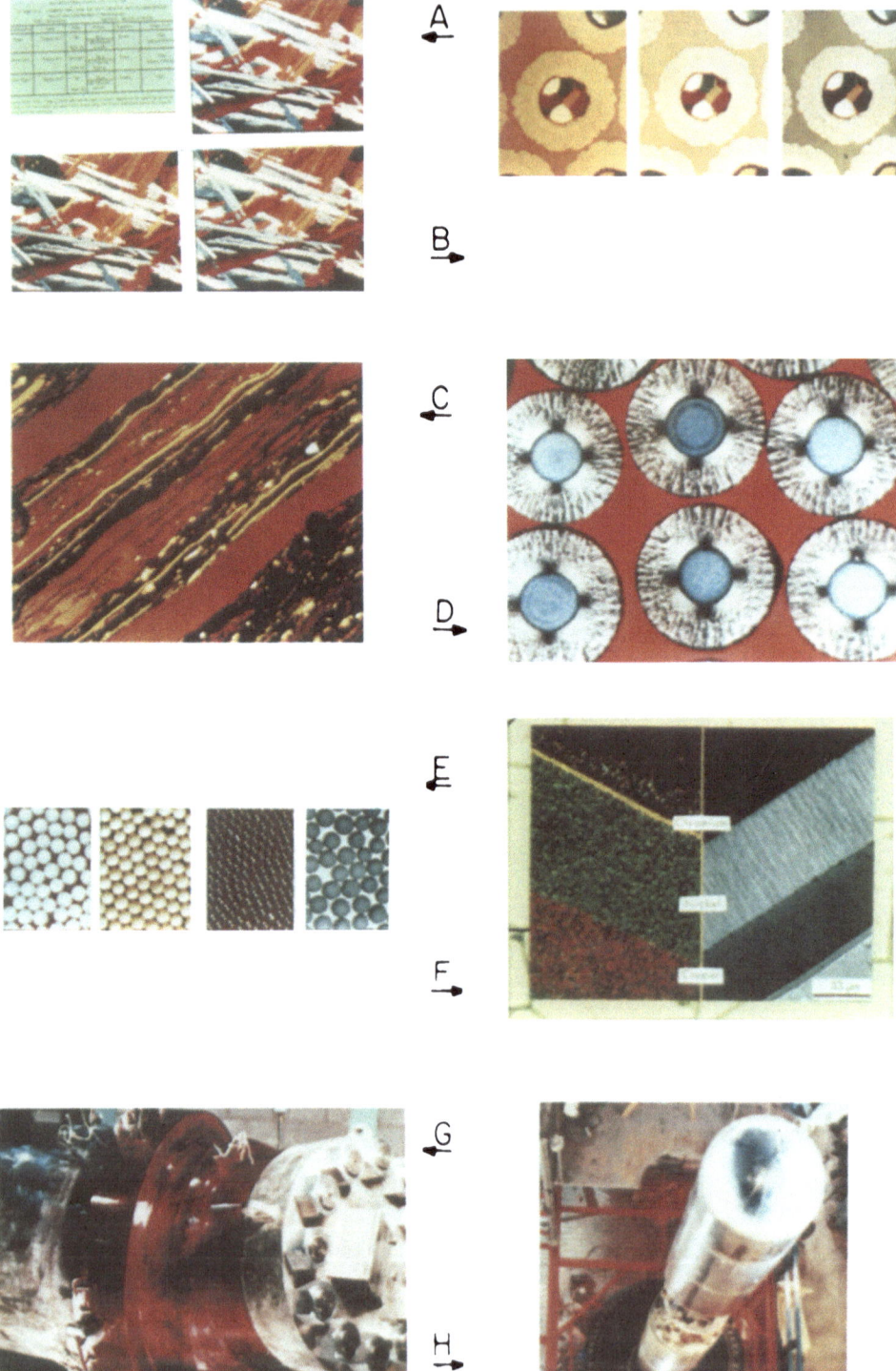

Figure 3

Plate A. This demonstrates the achievement of comparable color using different light sources and correcting filters. "Type B" film. A legible rendition of the table is found in the text as Table 4.

Plate B. This demonstrates the effect of three different filters on the overall color of a microstructure. The filters are, left to right, 85B, 85C, 82C. The 85C filter gives an image most nearly like the image observed through the microscope. The film used was a daylight emulsion and was exposed in polarized light with a Xenon lamp.

Plate C. Thin section of coal photographed in transmitted illumination on a metallograph. Since the microscope light used is rated at $3200°K$, the film used was a "type B" emulsion.

Plate D. Pyrolytic carbon coated nuclear fuel microspheres. This illustrates the effect of using a colored background to emphasize the microstructure of polished particles. Two exposures were made, one in polarized light with a carbon arc lamp and one with backlighting only. The film used was a "type B" emulsion.

Plate E. Ion exchange resin beads form the core of a fuel particle concept for use in high temperature gas-cooled power reactors. They are, left to right, as received, loaded and dried, carbonized, and pyro-coated. These were photographed on a Bausch and Lomb Research I metallograph using both external oblique sublighting and backlighting through color filters.

Plate F. Color-keyed composition map using microprobe X-ray displays. This composite shows the distribution of elements in the cross section of plated automobile bumper. The yellow is chromium, green is nickel, and red is copper. Photographed on "type B" emulsion film using $3200°K$ illumination.

Plate G. The water chest end of the steam generator in Fig. 10 with the dye for penetrant inspection applied.

Plate H. This is a Van der Graaff generator being assembled. The distance from the grating floor to the top of the cylinder is 2½ stories. This picture required a time exposure with fill-in flash exposures taken sequentially around the unit.

as an important tool in the study of these alloys. He worked out the electrolyte, voltage, and current densities so that one may consistently stain phases characteristic colors. For instance, the transformation phases in Zr–3% Ag are preferentially colored and this technique is used to follow the progress of transformation. This technique is also used to resolve and identify hybride in zirconium by its characteristic brilliant yellow color, shown in Fig. 1, color plate D.

The oxides, carbides, and nitrides of niobium and niobium alloys may be defined and identified by electrolytic anodization. Crouse [18] has reported a procedure for this identification and presented color photomicrographs of each of the phases.

A very simple and useful procedure for sigma phase identification in austenitic stainless steels is a dilute solution of NaOH used electrolytically (see appendix). An example of this is seen in Fig. 1, color plate E. The sigma phase is stained a bright orange-yellow in this Type 310 stainless steel. In instances where the phases are quite small, differentiation between carbide and sigma is only possible by using color.

A good example of grain contrast by electrolytic anodization is found in aluminum (see appendix). The film that is formed is a birefringent oxide layer with a surface topography that is related to the crystallographic orientation of the substrate. The explanation of this optical phenomenon has been presented [19,6]. The film displays strong response to plane-polarized light, and to achieve color a sensitive-tint plate is employed on the microscope so the plate revolves about the center of the optical axis[6]. Rotation of the plate produces colors that are truly spectacular, and slight variations of orientation within a grain due to deformation or sub grains are more clearly defined than with a fixed sensitive-tint plate or plane-polarized light. See Fig. 1, color plate F.

Other Film Deposition Methods

One technique involves the formation of a film in vacuum or normal atmosphere on a previously polished metallographic specimen. The deposit is truly an interference layer that results in constructive or destructive interference between the light reflecting from the film-specimen interface and the film-air interface. Pepperhoff[20] first reported this microstructural enhancement technique and specific application were reported by Stiegler and Gray[21].

One such application involves the vacuum deposition of titanium dioxide (high index of refraction) on a polished specimen. Various phases and components, such as mounting material in voids, oxides, and other corrosion products are shown in characteristic colors which enhance contrast and definition. A less complicated technique of applying a thin film of parlodian to produce microstructural contrasts has also been reported[22].

Another deposition procedure involves the use of a specifically designed contrasting chamber in which a selected gas is discharged against the specimen surface. This procedure, reported by Petzow[23], has a distinct advantage in allowing the metallographer to observe the formation of the film with a microscope through a window in the chamber.

Illumination Modes

Some materials in an as-polished condition have colorful microstructures when viewed in some illumination mode other than bright-field. Two of the most commonly available modes are sensitive-tint and polarized-light. A modification of the sensitive-tint mode uses a full-wave retardation plate that can be rotated out of its normally fixed position. This feature greatly increases the color contrast in weakly birefringent materials. All already have been mentioned above with regard to anodized aluminum.

In evaluating the performance of nuclear fuels, metallography has become a primary tool. There is usually no other good way to finally know just how a fuel is performing in irradiation but to polish and examine fuel sections. Fig. 1, color plate G, is a fuel microscope coated with pyrolytic carbon and embedded in a carbon matrix. The yellow phase is graphite that has been formed by evaporation of the coating from the hot side of the fuel and deposition on the cooler side. This structure was photographed in sensitive-tint illumination since it was difficult to see in bright-field or polarized-light.

Reactor-grade graphite is relatively nondescript in appearance when viewed in bright-field or polarized-light illumination. However, using the rotatable sensitive-tint illumination to increase color contrast produces considerable structure emphasis. Fig. 1, color plate H, is "needle-coke" graphite, so called because of the elongated nature of its filler coke particles. The materials engineer uses this optical technique to evaluate his product[24].

Other materials that show colorful microstructures in the as-polished condition are europium silicide and beryllium boride. Some others that display enhanced color contrast from an etched condition due to illumination mode are uranium and thorium carbides.

LIGHT SOURCES AND FILTERS

Metallographs and microscopes come equipped with a variety of light sources from tungsten-ribbon filaments to xenon lamps. Table 2 lists the light sources generally available and the filters to use with the two types of color photographic emulsions. Any of these light sources can be used to make acceptable color exposures providing the proper filters are chosen. The ribbon filament, because of its low color temperature, may be marginally suitable, but all the others are excellent. One light source occasionally found on metallographs is the mercury vapor lamp. This light is unsuitable for color because of its intense green color and discontinuous spectrum. Some wavelengths are completely missing.

Additional filtering in the form of an infrared cut-off filter and a sodium nitrite solution cell are required for both the carbon arc and the xenon lamp. This is to remove infrared and ultraviolet radiation, both of which, although invisible, affect the color emulsion. Color plate A of Fig. 3 demonstrates the possibility of accurate color reproduction by proper filter selection for different light sources. (See Table 3 for a legible rendition of the table shown in this plate).

Table 2

LIGHT SOURCES AND FILTERS FOR COLOR METALLOGRAPHY

Light Source	Filter	
	Tungsten Emulsion	Daylight Emulsion
Ribbon Filament (2900° K)	82B	80B + 82C
High Wattage Temperature (3200° K) 300-1000 W	none	80B + 82A
Zirconium Arc (3200° K)	none	80B - 82A
[a]Carbon Arc, 4.5 amp (3650° K)	81C	80C - 82
[a]Carbon Arc, 10 amp (3800° K)	81D	80C
[a]Xenon Lamp, 25 amp (6700° K)	81A + 85B	81A

[a]Use No. 301 Infrared Cutoff Filter and Sodium Nitrite (3/4%) Solution in a Water Cell.

Table 3

MICROSTRUCTURES IN COLOR

10 AMP D.C. CARBON ARC AND 450 WATT XENON ARC ILLUMINATION
Bausch and Lomb Research I Metallograph

Film: Ektachrome   Specimen: Beryllium Boride

Identification	Lamp	Film	Filters	Exposure	Water Cell
Right	Carbon Arc*	"B" ASA 25	81 C / 301 Infrared cut-off	18 sec	0.75% Sodium Nitrite
Lower Left	Xenon Arc	Daylight ASA 50	82 A / 0.3 Neutral Density	4 sec	0.75% Sodium Nitrite
Lower Right	Xenon Arc*	"B" ASA 25	85 B / 301 Infrared cut-off	6 sec	H_2O

*Ektacolor Prof. Film, Type L (ASA 64) can be used to obtain negatives for positive prints or print film with the carbon arc and xenon lamps using the filters indicated above for type B film and half the exposure.

Table 4 is a copy of a Bureau of Standards filter nomograph that aids in the selection of the best filter for a given light source and emulsion. The filters suggested may not be exactly right for accurate color reproduction, but they will always be close enough for one to make intelligent changes to achieve accuracy.

To illustrate the effect of different filters on the finial image, color plate B of Fig. 3 shows exposure through three different filters. The exposure through the 85C filter was deemed to be nearest to the color seen in the microscope. Some experimentation may be necessary to finally arrive at a satisfactory color balance.

Exposure determination in photomicrography generally is best done empirically. The authors are unaware of any commercial exposure meter of sufficient sensitivity to use in polarized-light illumination, so a practical guide has been developed using Polaroid P/N, Type 55 film. It has the same film speed as Ektacolor, Type L, which is often used for making color prints. Other materials are exposed in reference to the Type 55 based on comparative film speeds, i.e., a film with twice the speed (ASA rating) would take half the exposure time.

No discussion of color photography would be complete without some reference to Polacolor film. This material is satisfactory for a limited number of prints in the 4 inch by 5 inch format. In cases where larger prints and many copies are needed, one must use a negative or positive transparency material. The "keeping quality" of Polacolor is very good, but the red color saturation leaves something to be desired. Polacolor II, as of this writing, has not been evaluated by the authors.

COMMERCIAL FILMS COMPARISON

Not all metallography labs are equipped to do all the processing of color films, and because of this, miss out on the advantages of using it. With all the recent improvements in films and projection equipment, more and more people are using 35 mm slides in preference to the 3¼ inch x 4 inch glass lantern slides. Excellent 35 mm cameras are available and many microscopes and metallographs come equipped in the 35 mm format. It is possible, without much expense, to adapt this format to older equipment as well.

Fig. 4 shows a Bausch and Lomb Research I metallograph adapted to the 35 mm format. Any miniature camera body can be attached to the projection lens in this manner, although a single-lens-reflex camera has the obvious advantages of allowing accurate focussing. This sort of arrangement also provides a means of determining exposure by inserting a commercial meter in the eyepiece tube. Many such meters can be bought with eyepiece adapters. Calibration for the particular instrument-emulsion combination should be straightforward.

Although the 35 mm camera equipment and procedures described above are quite satisfactory, a modest step-up in camera selection is very worthwhile. Electronic single-lens-reflex cameras, such as the Minolta XK, offer automatic exposure and an electronically-timed shutter. The adaptation of a camera of this type to a microscope has been reported [25]. Almost any lens system, including a bench or console type metallograph, can be mated to the Minolta XK camera body. This union, which replaces the standard

Table 4

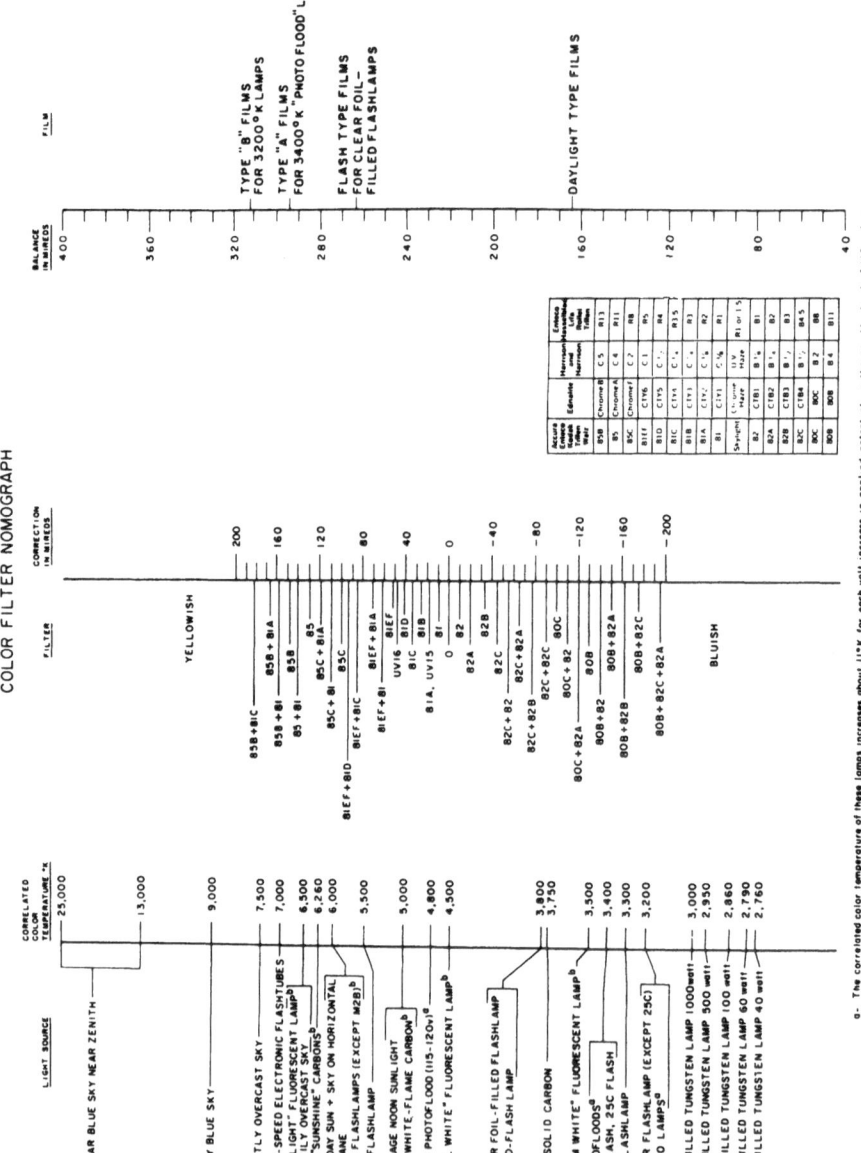

COLOR FILTER NOMOGRAPH

Fig. 4. Single lens reflex 35 mm camera and exposure meter mounted on Bausch and Lomb Research I metallograph.

camera lens, does not interfere with the original functions of automatic exposure and electronically timed shutter. If a 35 mm camera must be purchased for a microscope attachment, we recommend this much improved system. Figs. 5 and 6 show the Minolta XK attached to Bausch and Lomb Research I and II metallographs, respectively.

In an effort to be fair in the consideration of color photographic film, a comparison of eight different transparency films was made[26]. Fig. 7 compares these films with each other and with Eastman Kodak Ektachrome X, which has been used as a standard.

The subject for this comparison was as-polished beryllium boride viewed in rotated-sensitive-tint illumination. The microscope and camera were the set up as shown in Fig. 4. Each different film was sent to recommended commercial processors and a record kept of the turn-around time. Table 5 summarizes the results of the comparison with comments about the general appearance of each film.

Efforts were made to bracket the proper exposure, but in the case of the GAF film, the best exposure was somewhat under the optimum. The extremely high film speed (ASA 500) caused a miscalculation in shutter speed. A mental projection of how a properly exposed image would look leads one to feel that it would compare favorably with the standard.

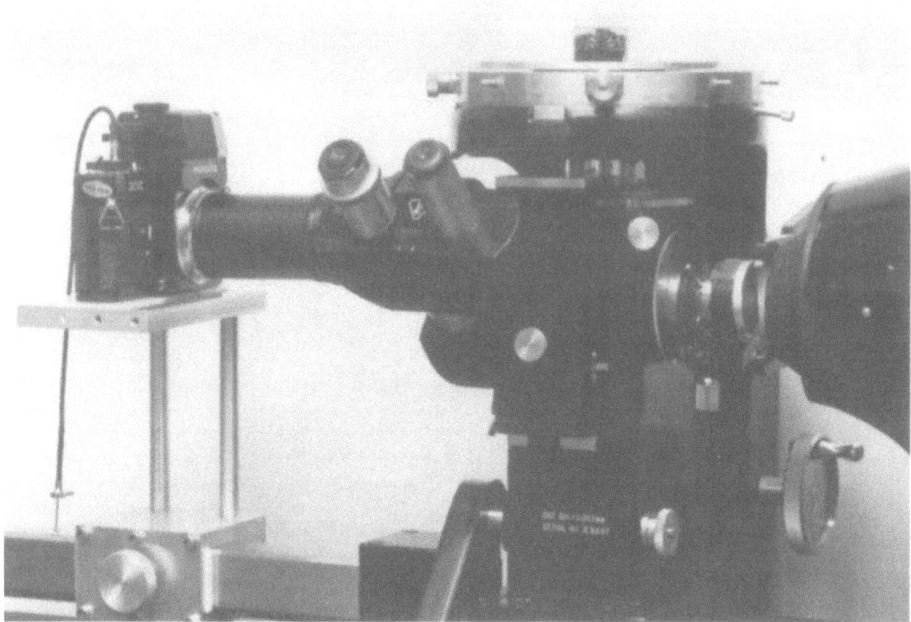

Fig. 5. Camera with electronic shutter for automatic exposure control mounted on Bausch and Lomb Research II metallograph.

A conclusion that can be drawn from this study is that a small lab may produce perfectly acceptable 35 mm slides of photomicrographs as simply as one might handle his own personal photography. Commercial processing of transparency films is highly reliable and relatively rapid, and accelerated service is available from some photo labs with 48 hr service sometimes available. The choice of film is up to the preference of each lab.

SPECIAL LIGHTING EFFECTS

Occasionally, it becomes necessary to resort to auxiliary or special lighting to adequately record a color microstructure. One such special arrangement is shown in Fig. 8. This is used to provide transmitted illumination for such things as petrographic thin sections.

The constantly increasing crisis in energy has brought a renewed interest in coal and coal research. Belying its generally commonplace external appearance, a thin section of coal exhibits a surprising amount of color. Fig. 3, color plate C, is coal photographed using the technique illustrated in Fig. 8. The layered microstructure reflects the manner in which the original organic material was laid down millions of years ago. The yellow material is said to be spores and pollen which have been highly compressed.

Color plate D of Fig. 3 demonstrates another way in which the backlighting feature is used. In this case, the red background was provided to accentuate the blue and silver

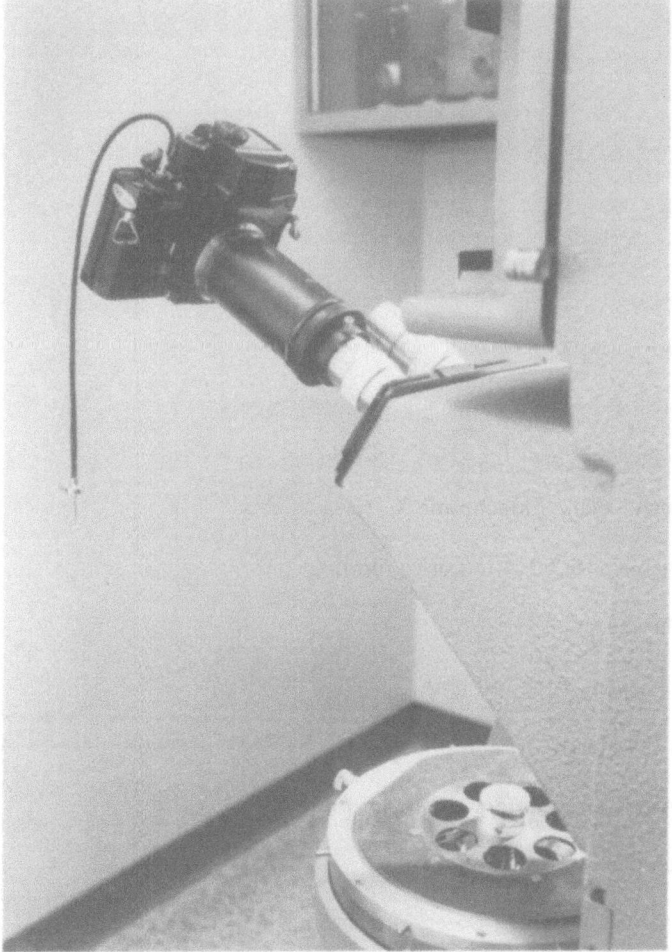

Fig. 6. Camera with electronic shutter for automatic exposure control mounted on Bausch and Lomb Research II metallograph.

of the polished-fuel microspheres. It was obtained by casting a layer of red paint in the specimen mount at a level above the microspheres so as to be out of focus. The microspheres were photographed in polarized light and then an additional exposure was made using only the backlight (Fig. 8) to emphasize the red background.

Currently, one of the ways of producing nuclear fuel microspheres is to start with ion exchange resin beads, load them with uranium, convert them to carbide, and then coat them with pyrolytic carbon. After each step the microspheres display a different texture and color as shown in Fig. 3, color plate E. The significant part of this photomacrography is that the photos were made on a Bausch and Lomb metallograph using

Figure 7

Plate A. Eastman Kodak Ektachrome X.

Plate B. Eastman Kodak 5254 (negative film).

Plate C. Fujichrome.

Plate D. GAF (General Archive Film).

Plate E. Focal (K—Mart).

Plate F. Eastman Kodak Photomicrography Film.

Plate G. Agfachrome.

Plate H. Eastman Kodak High—Speed Ektachrome.

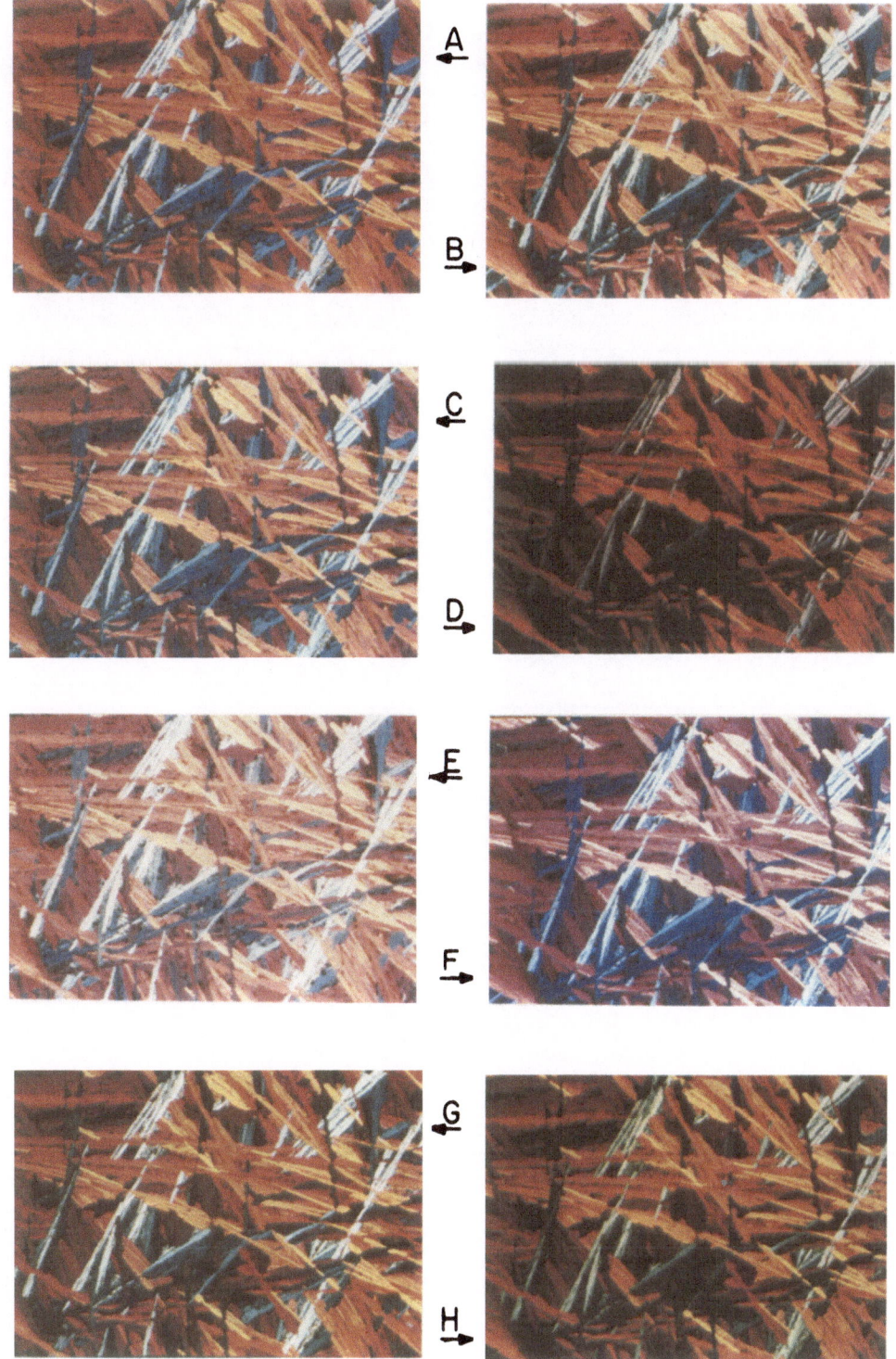

Table 5

COMPARISON OF EIGHT FILMS FOR METALLOGRAPHIC APPLICATIONS
Xenon Lamp - 25 amp.   Bausch and Lomb Research 1 Metallograph
Projected Magnifications: 30 X, 52 X, 82 X

Film	ASA	Exposure (sec)	Filter	Processing	Remarks
E. K. 5254 (neg) T	100	1/15	301 Inf. Red. water cell/Ht. Absorb.	Hollywood, CA 8 days	Medium contrast Color fairly well balanced Some green evident
Fujichrome (pos.) D	100	1/30	82 A + .75% sodium nitrite sol. with heat absorb. filter	Solar Chicago 11 days	Low-Medium contrast Color fairly well balanced Blue hue
GAF (pos.) D	500	1/125	"	GAF Philadelphia 20 days	Medium contrast
Focal (pos.) D	400	1/60	"	GAF Philadelphia 16 days	Low contrast
E. K. Photomicrography (pos.) D	16	1/2	"	Eastman Kodak Atlanta 7 days	High contrast Pronounced blue, red, yellow
E. K. Ektachrome X (pos.) D	64	1/15	"	Eastman Kodak Atlanta 7 days	Medium contrast Pronounced blue
Agfachrome (pos.) D	64	1/15	"	Honeywell Flushing, NY 9 days	Medium contrast Colors fairly well balanced
E. K. Hi Speed Ektachrome (pos.) D	160	1/60	"	Eastman Kodak Atlanta 9 days	Medium contrast Pronounced green

D - Daylight; T - Tungsten

a Zeiss 60 mm lens with an oblique lighting arrangement shown in Fig. 9. In addition, light from above the stage through a color filter chosen to accent the natural color of the spheres formed the backgrounds. This lighting technique allows the choice of any color background desired to enhance color contrast or hues.

A technique Yakowitz[1] calls color separation metallography is sometimes used to present x-ray element mapping data obtained from an electron microprobe or scanning electron microscope. As each element in a sample is displayed on the instrument cathode ray tube (CRT), it is photographed on Polaroid 46L projection film which is rated at ASA 800. These are then arranged so that they may be photographed on a single piece of color film one at a time through color filters arbitrarily assigned to each element. Such a photographic setup is shown in Fig. 10. Normally the paper mask would be black, but for the sake of clarity it is shown here as white. Each piece of film is taped along one edge so that one at a time they may be photographed through the hole (6 cm x 8.5 cm) in the mask. The resultant color composite contains all the elements in their assigned colors. Care must be taken to have the images in registry when the films are taped down.

Color plate F, Fig. 3, is an example of this technique. This is a cross section of a chrome-plated automobile bumper and shows how, in a single photograph, the element distribution is related to the microstructure. Polacolor film is a good material to use for the color composite, although any other color film could be used.

Fig. 8. An arrangement for photographing transparent samples, such as petrographic thin sections, on a Bausch and Lomb Research I metallograph.

STUDIO AND FIELD PHOTOGRAPHY

As a natural extension of the photography in a metallography lab, studio and field photography also often are required. If standard photographic equipment such as 35 mm and view cameras are available for photomicrography, eventually requests will be made for color slides of copy work and industrial objects. A relatively inexpensive home-made photocopy arrangement is shown in Fig. 11. A vacuum easel was built by tying a shallow box covered with ¼ inch pegboard into the building exhaust system. The light bars are $3200^\circ K$ floodlamps on folding arms which allow them to be moved out of the way when not in use. With backdrop material available, this lighting system also can be used for still photography.

Fig. 9. Oblique-and transmitted-light arrangement for microsphere photography on a Bausch and Lomb Research I metallograph.

Field photography as related to metallography frequently involves taking pictures of items that eventually will produce metallographic specimens. The purpose for taking such pictures is for documentation and inclusion in reports. Color photography usually is not mandatory, but occasionally is indicated to emphasize some specific feature. Usually, the expense of reproduction precludes the inclusion of color photographs in reports, but for small numbers of informational handouts color prints serve a worthwhile purpose.

Postmortem examination of large industrial components demands complete photographic documentation every step of the way. Fig. 12 is a nuclear power station steam generator being put in place for disassembly. This would be the first of a series of photographs that would record each operation and might or might not be in color.

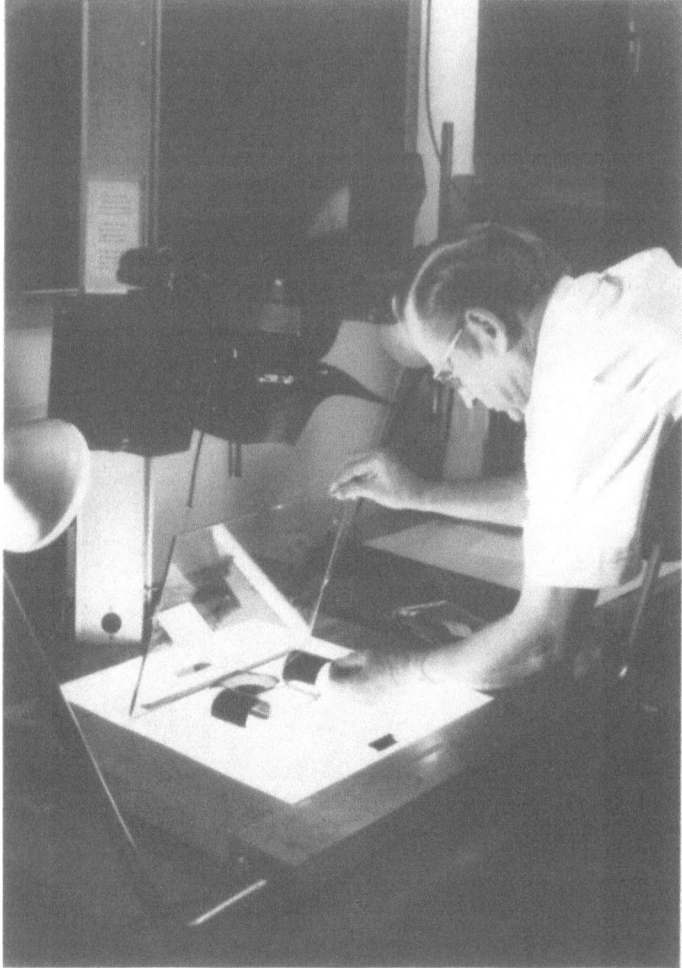

Fig. 10. Set-up for color composite photography.

One of the first tests performed in this particular postmortem examination was a dye penetrant inspection. Fig. 3, color plate G, shows the water chest of the generator with the dye applied. The visual impact of this picture is greatly enhanced by color photography.

Color plate H of Fig. 3 shows a Van der Graaff generator being assembled. The unit is two and one-half stories tall and required a time exposure featuring sequential flashes from an electronic flash unit at various points in the room.

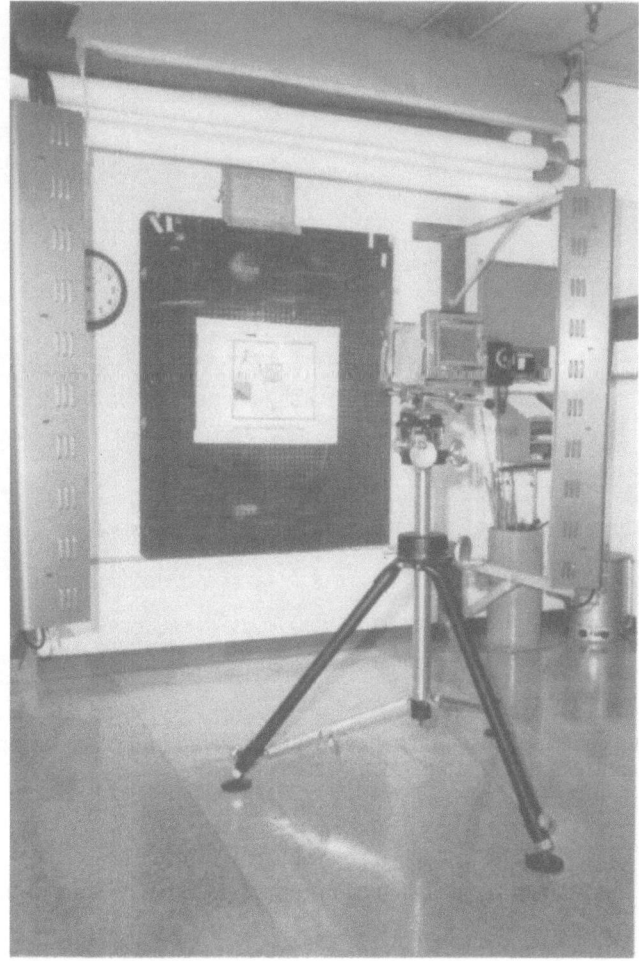

Fig. 11. Arrangement for photocopying employing a vacuum easel and 3200°K light bars.

CONCLUSIONS

Color photography as applied in a metallography laboratory has moved out of the realm of a laboratory curiosity into one of considerable utility. We have come to recognize that in some instances there is just no substitute for recording our observations in color. The increased microstructural contrast and analytical information provided by interference films on metallographic specimens demand it. The current state of photographic technology makes it possible for even small labs to indulge in color to a certain extent. Good cameras and speedy, reliable commercial film processing are the tools, and they are available. They are not unbearably expensive and the end results are worth the little added effort.

Fig. 12. Steam generator (heat exchanger) being lowered into place for preassembly.

APPENDIX

Metallographic conditions for producing color in several materials noted in the text:

Yttrium

Immerse in equal parts of acetic, nitric, and phosphoric acids. Examination in bright-field illumination.

Molybdenum

Immerse in 70 parts water, 20 parts 30% hydrogen peroxide and 10 parts sulfuric acid . Examine in bright-field illumination.

Uranium Carbide

Immerse in equal parts acetic acid, nitric acid, and water. Examine in bright-field or polarized illumination.

Sigma Phase in Type 310 Stainless Steel

Etch electrolytically in saturated sodium hydroxide at 1.5 volts for 2 sec. Examine in bright-field illumination.

Aluminum

Electropolish from 600 grit paper in 40 parts fluoroboric acid, 50 parts water at 25–30 volts, preferably in an automatic system such as Disa Electropol.

Electroetch with solution of 23 parts above solution and 977 parts water— same conditions. Examine in polarized or rotated-sensitive-tint illumination.

REFERENCES

1. H. Yakowitz, "Some Uses of Color in Metallography", *Applications of Modern Metallographic Techniques,* ASTM STP 480, American Society for Testing and Materials, 49–66 (1970).

2. F.W. Young, J.V. Cathcart, and A.T. Gwathmey, "The Rates of Oxidation of Several Faces of a Single Crystal of Copper as Determined with Elliptically Polarized Light," *Acta Met* 4(2), 145–163 (1956).

3. J.V. Cathcart, G.F. Peterson, and C.J. Sparks, Jr., "Lattice Disregistry in Very Thin Oxide Films on Copper," *Mem. Sci. Rev. Met.* 62, 11–16, (May, 1965).

4. J.V. Cathcart, G.F. Peterson, and C.J. Sparks, "Oxidation Rate and Oxide Structural Defects," *Surfaces and Interfaces Chemical and Physical Characteristics,* Burke, Reed, Weiss Ed., Syracuse University Press, Syracuse, N.Y. (1967).

5. R.E. Pawel, J.V. Cathcart, and J.J. Campbell, "Oxide Platelet Formation in Tantalum Single Crystals," *Acta Met* 10(2) 149–160 (1962).

6. R.J. Gray, R.S. Crouse, and B.C. Leslie, "Decorative Etching," *Metallographic Specimen Preparation,* Ed. by McCall and Mueller, Plenum Publishing Co., New York, N.Y., 179–206 (1975).

7. R.J. Gray, "Present Status of Metallography," *Fifty Years of Progress in Metallographic Techniques,* STP 430; ASTM, Philadelphia, PA, (April, 1968).

8. R.J. Gray, W.C. Thurber, and C.K. H. DuBose, "Preparation of Arc-Melted Uranium Carbides," *Metal Progress,* 74, 65–70 (July, 1964).

9. R.W. McClung, E.S. Bomar, and R.J. Gray, "Evaluating Coated Particles of Nuclear Fuel," *Metal Progress,* 86(1), 90–93 (July, 1964).

10. E. Beraha, *Journal of the Iron and Steel Inst.,* Vol. 202, 696 (1964).

11. E. Beraha, *Journal of the Iron and Steel Inst.,* Vol. 203, 454 (1965).

12. E. Beraha, *Journal of the Iron and Steel Inst.,* Vol. 204, 248 (1966).

13. E. Beraha, *Metal Progress,* 9(3), 135 (1966).

14. E. Beraha, *Journal of the Iron and Steel Inst.,* Vol. 205, 866 (1967).

15. E. Beraha, *Praktische Metallographie,* 4(8), 416 (1967).

16. E. Beraha, *Praktische Metallographie,* 5(9), 501 (1968).

17. M.L. Picklesimer, "Anodizing as a Metallographic Technique for Zirconium Base Alloys," ORNL–2296, Clearing House for Federal Scientific and Technical Information (1957).

18. R.S. Crouse, "Identification of Carbides, Nitrides, and Oxides, Niobium, and Niobium Alloys by Anodic Staining," ORNL–3821 (July, 1965).

19. E.C.W. Pergman, ' The Examination of Metal Surfaces," *Polarized Light in Metallography,* ed. by G.K.T. Conn and F.J. Bradshaw, Butterwirths Scientific Publication, 70–89 (1952).

20. W. Pepperhoff, ' Sickbarmachung on Gefugestrukturen durch Interferentz-Aufdampfschichten," *Des Naturwissenschaften* 16, 375 (1960).

21. J.O. Stiegler and R.J. Gray, "Microstructural Discrimination by Deposition of Surface Films," 11–17 in *Advances in Metallography* (ed. by R.J. Jackson and A.E. Calabra), Technical Papers of the 20th Metallographic Conference, The Dow Chemical Company, RFP–658 (Oct., 1966).

22. R.E. Staub, Jr., and J.L. McCall, "Increasing the Microscopic Contrast of Phases with Similar Reflectivities," *Metallography* 1(1), 153–155 (Sept., 1968).

23. G. Petzow and H.E. Exner, "Developments and Trends in Preparation Techniques for Metallographic Samples," *Microstructural Sci.* Vol. III (ed. by P.M. French, R.J. Gray, and J.L. McCall), Elsevier Publishing Co., Inc. New York, N.Y. (1975).

24. W.H. Cook, M.D. Allen, B.C. Leslie, and R.J. Gray, "High-Resolution Optical Microscopy of Carbon and Graphite," *Microstructural Sci.* Vol. III (ed. by P.M. French, R.J. Gray, and J.L. McCall), American Elsevier Publishing Co., Inc., New York, N.Y. (1975).

25. William Bohrer, "A New 35 mm Application: Mating a Minolta XK with a Balphot II, " *Industrial Photography* 24(10), 40–41 (Oct., 1975).

26. Alfred M. Gordon and Hiroshi Kimata, "Which 35 mm Color Slide Film is Better?," *Modern Photography* 40(4), 80–85, 156, 164, 166 (April, 1976).

MEASURING WITH THE OPTICAL MICROSCOPE

M. NAHMMACHER *

INTRODUCTION

The microscope was conceived as an instrument for seeing small things larger, as the name implies. But I would think that it took only a few hours before the user of the first microscope asked the question, "How small is it"? And with this question, a metamorphosis of the microscope from scope to meter began.

A thorough search of the literature in preparation for this presentation led me to the journal of the New York Microscopical Society. On Page 211 of Volume 1, Nov. 1885, it reads, "No microscope is complete unless equipped with ample and accurate means of micrometric measurement, and no man who does not fully understand the use of those means is entitled to be called a professional microscopist. He who possesses this accomplishment, has facilities for adding valuable material to the stock of human knowledge."

Before discussing "Micrometric Measurements" in detail, the question arises, which specimen parameters can one measure with a microscope or, in other words, which measurable information is contained in the microscopical image? The image can be said to consist of a large but finite number of image points. Their number and size are determined by the numerical aperture and magnification of the objective. These image points carry information about the specimen, and one must ask for the nature of this information.

1. LOCATION : Each image point has a fixed location which can be characterized by its coordinates within the image. From the measurement of the distance between two such image points and the determination of their angular relationship within the image one arrives at all measurements in the two dimensions of the specimen at right angle to the axis of the microscope. The results are the quantification of size, area, shape, perimeter, spacial relationships and, finally, the recognition of patterns within the image.

2. LIGHT INTENSITY: An image point has a certain brightness which can be measured by means of a photodetector. Such a measurement of light intensity

*E. Leitz, Inc., Rockleigh, New Jersey 07047

leads to the quantification of absorption, reflection, remission and fluorescence.

3. PHASE RELATIONSHIPS: Image points also carry phase information and the phase relationship between two image points is measurable by means of an Interferometer. The measurement of such phase shifts accurately determines distances along the optical axis of the microscope within the specimen. Using reflected light, one can quantify a surface profile or simple obtain the height or depth of a surface structure. In transmitted light, it is possible to measure the refractive index and the thickness of a transparent specimen because both parameters affect the phase relationship of image points.

4. POLARIZATION: The light associated with an image point is in a certain state of polarization which is a function of optical specimen properties, such as, birefringence, bereflectance, and optical orientation. Related properties are the different forms of dichroism, optical activity and optical rotatory dispersion. The state of polarization can be measured very accurately in a polarizing microscope which permits the characterization or even the identification of a microscopic specimen. The strain in an isotropic material and the molecular order in a polymer are related specimen parameters which can be quantified in the same manner.

Now one must ask: Where in the optical train of the microscope is the most suitable place to carry out a measurement? The answer to this question depends on the type of measurement and on the required accuracy. The intermediary image produced by the objective and located within the eyepiece is most frequently used. However, here the magnification is relatively low which can limit the measuring accuracy and presents a serious disadvantage in microscope photometry because very small microscopic structures cannot be properly isolated for the measurement. In these cases, one chooses a second intermediary image with higher magnification. This may be the photographic plane in a camera system or the plane of the measuring window in a microscope photometer. Smaller measuring sites and improved repeatability result.

For the sake of completeness, it should be mentioned that measurements are not necessarily carried out only in image planes. The back focal plane of the microscope objective is used for conoscopy in polarized light microscopy in order to measure, for example, the optical orientation of the specimen with respect to the optical axis of the microscope. It is also used for Fourier analysis.

The measurements with a light microscope can be categorized into four groups: 1. Measurements in the Two Dimensions of the Image Plane; 2. Measurements along the Axis of the Microscope; 3. Measurements of Light Intensity; 4. Measurements of Temperature.

MEASUREMENTS IN THE TWO DIMENSIONS OF THE IMAGE

Linear Measurements

In the simplest case, a distance in the image is measured by means of a fixed eyepiece micrometer. Since the magnification marked on the objective may not be accurate, it is necessary to determine the micrometer value with a stage micrometer. In this way, the relationship between a unit of the micrometer with the corresponding distance in the specimen plane is established. Filar micrometer or image shearing eyepieces offer greatly improved reading accuracy. A still further improvement is a television micrometer with a digital read-out. Here, the microscope is not touched during the measurement and the relationship between the moving line and the image is, therefore, absolutely stable.

The methods so far discussed are limited to the measurement of structures which are smaller than the field of view of the microscope. This is not the case with micrometer spindle stages. Spindles with a range of two inches and a reading accuracy of 10/1,000,000 ± 15/1,000,000" are available.

The accuracy of the linear measurement is closely related to the resolution of the microscope objective because it limits the ability of the operator to accurately locate the border of the structure to be measured. Therefore, high aperture bright-field microscope is best suited for linear measurements. It is also helpful to use empty magnification for reasons of convenience and accuracy. Contrast enhancing techniques, such as, phase-contrast or differential interference contrast do not improve the repeatability and accuracy of the measurement, but frequently lead to measuring errors. This is different with the television micrometer. By processing the video signal, borders of image structures can be dramatically sharpened and improved repeatability results as shown in Fig. 1.

1. Subject Prior to Processing

2. With Sharpening, no Noise Suppression

3. With Sharpening, and Noise Suppression

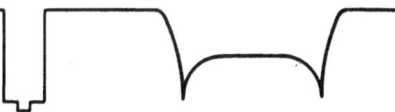

Figure 1. Processing of a video signal (courtesy of ITP)

Hardness Testing

An indirect application of linear measurement is the determination of the hardness of a material with a micro-hardness tester. The diagonals of Knoop or Vickers indentations are measured with a filar micrometer eyepiece which is calibrated in micrometer units. Here additional factors enter into the accuracy of the measurement, factors which affect the correlation between the size of the indentation and the hardness of the material. For example, the depth of the indentation must be kept small in comparison to the thickness of the specimen and the indentation should not come too close to the periphery of the tested structure. (See Fig. 2.)

Angular Measurements

Angles within the specimen can be measured directly by means of a goniometer or protractor eyepiece containing a rotatable reticle illustrated in Fig. 3. These eyepieces usually have a reading accuracy of one minute. Ordinary rotating stages are not as accurate and can be read only to one tenth degree with the aid of a venier.

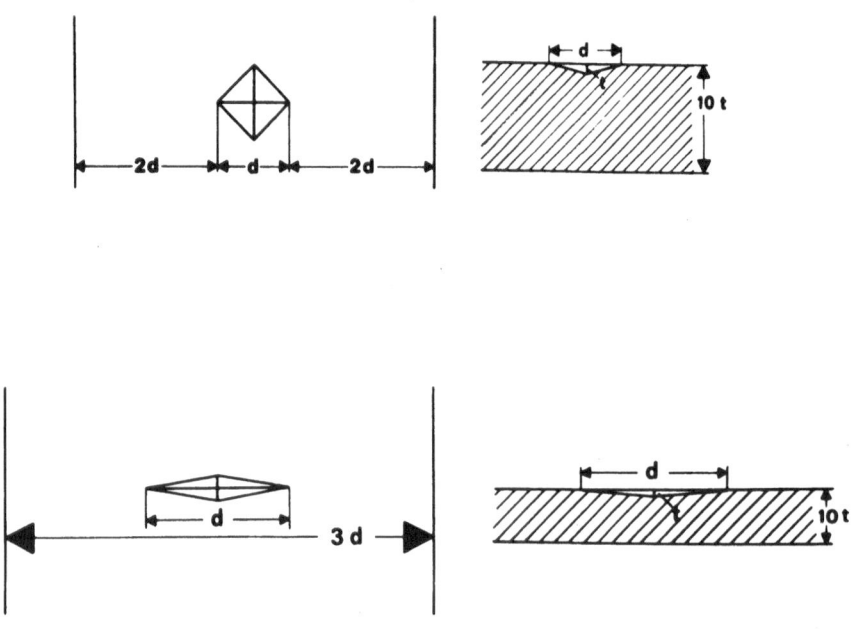

Figure 2. Vickers and Knoop hardness indentations.

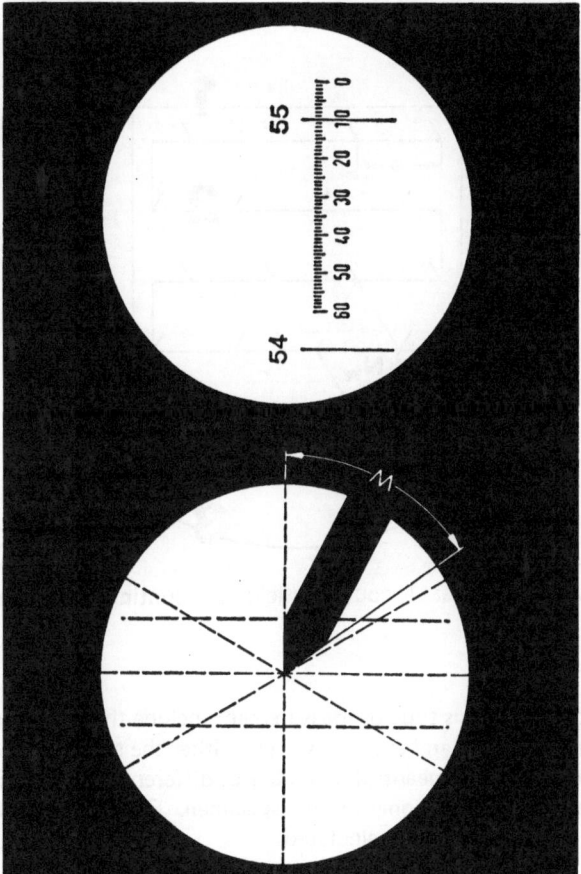

Figure 3. Goniometer or Protractor Eyepiece

Area Related Measurements

The percentage area of a specimen component can be measured by means of point counting reticles which are placed into the intermediary image plane in the eyepiece. They consist of an array of point targets. Those targets which are superimposed over the specimen component are counted. From this count, one calculates the percentage area and specimen parameters of similar nature.

The Blaschke reticle may serve as an example as shown in Fig. 4. Three different counting values are determined.
1. The number of cross-sections within the rectangular measuring field (Q).
2. The number of hits (T).
3. The number of grain boundary intersections with the counting circle (S).

From these values, one can calculate the grain and structure quantities which are listed in Table 1.

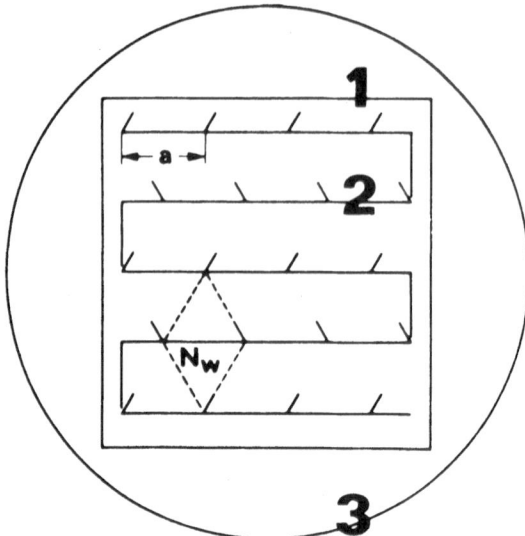

Figure 4. Blaschke Reticle: 1. counting field; 2. counting point reticle; 3. counting circle.

Other approaches to this type of measurement include the use of point counting stages or integrating stages and eyepieces. Here either the stage or a target within the eyepiece are moved by means of a number of different spindles. Each spindle is assigned to a different component in the specimen. From the total travel of each spindle, the percentage areas are calculated.

All these methods are extremely time consuming and tedious. This is the reason why they are employed only in relatively few fields of application and even these only to a limited degree. For many other applications, especially in industry, they have proved to be rather impractical. Only when the first television image analyzers became commercially available a few years ago did such measurements become widely used for material analysis.

Television Image Analysis

The television raster becomes the reticle for point counting. The large number of raster points insures high sampling resolution and measuring accuracy, and the speed of the television scanning system makes it now a practical method which can be employed in a wide variety of fields. Commercial instruments of the first generation all permit the determination of those image parameters mentioned above within reasonable time limits, but their capabilities to truly characterize an image are limited due to the fact that they are linear analyzers which scan only in the direction of the scanning line.

TABLE 1
Formulae for the calculation of the grain—and structure—
specific quantities from the counting quantities T,Q,S.

Designation of the counting quantity		Formula and dimension when a in cm	Formulae for practical calculation
$F_\%$ $V_\%$	Area percentage in the plane of section Volume percentage	$F_\% = V_\% = 100 \dfrac{\text{Number of hits}}{\text{Number of counting points}}$ [dimensionless]	$F_\% = V_\% = \dfrac{5}{n} T$
Z	Number of the grain cross sections per unit area	$Z = \dfrac{\text{Number of cross sections}}{\text{Area of the counting field}}$ [cm^{-2}]	$Z = \dfrac{Q}{17.32\, a^2 n}$
F_{qfl}	Mean area of the grain cross sections	$F_{qfl} = \dfrac{\sqrt{3}}{2} a^2 \dfrac{T}{Q}$ [cm^2] (according to HENNIG, 1958)	$F_{qfl} = 0.866\, a^2 \dfrac{T}{Q}$
Derived from the mean area of the grain cross sections (F_{qfl}) — b_{qfl}	mean side length of square grain cross sections	$b_{qfl} = \sqrt{F_{qfl}}$ [cm]	$b = 0.9306\, a \sqrt{\dfrac{T}{Q}}$
d_{qfl}	mean linear extent of the grain cross sections	$d_{qfl} = 2 \sqrt{\dfrac{F_{qfl}}{\pi}}$ [cm]	$d_{qfl} = 1.050\, a \sqrt{\dfrac{T}{Q}}$
D_{qfl}	mean grain diameter	$D_{qfl} = \dfrac{4}{\pi} d_{qfl}$ [cm]	$D_{qfl} = 1.337\, a \sqrt{\dfrac{T}{Q}}$
V_{qfl}	mean grain volume	$V_{qfl} = \dfrac{4}{3} \pi \left(\dfrac{D_{qfl}}{2}\right)^3$ [cm^3]	$V_{qfl} = 0.5236\, D^3_{qfl}$ $= 1.251\, a^3 \dfrac{T}{Q} \sqrt{\dfrac{T}{Q}}$
O_i	Internal specific area of a phase, referred to the volume unit of the complete system (e.g. area of the quartz grains in 1cc of rock)	$O_i = 2 \dfrac{\text{Number of grain boundary intersections}}{\text{length of intersection line}}$ [cm^2 per cm^3 = cm^{-1} complete system]	$O_i = \dfrac{S}{10\, an}$
O_{sp}	Specific area of a phase (e.g. S) referred to the volume unit of this phase (e.g. area of a cc of quartz in grains)	$O_{sp} = \dfrac{\text{Internal specific area}}{\text{Volume percentage}}$ [cm^{-1} phase]	$O_{sp} = 2 \dfrac{S}{aT}$
Derived from the specific surface area (O_{sp}) — b_{spo}	mean edge length of cubic equivalent grains	$b_{spo} = D_{spo} = \dfrac{6}{\text{specific area}}$ [cm]	$b_{spo} = D_{spo} = 3 \dfrac{aT}{S}$
D_{spo}	mean diameter of spherical equivalent grains		
V_{spo}	mean volume of spherical equivalent grains	$V_{spo} = \dfrac{4}{3} \pi \left(\dfrac{D_{spo}}{2}\right)^3$ [cm^3]	$V_{spo} = 1.523\, D^3_{spo}$ $= 14.137 \dfrac{a^3 T^3}{S^3}$
f	form factor	$f = \dfrac{V_{qfl}}{V_{spo}}$ [dimensionless]	$f = \left(\dfrac{D_{qfl}}{D_{spo}}\right)^3$ $= 0.0885 \left(\dfrac{S}{\sqrt{QT}}\right)^3$

Recently, the second generation of image analyzers has become available which not only permits a two-dimensional scan, but have also extended capabilities for linear analysis.

Linear Scanning Elements. Instead of one point, any group of points can be used as scanning element. Depending on the nature of the scanning element, for example, a cumulative or absolute cord length distribution can be established, or the probability with which a segment of a given length is found in the image can be determined.

Covariance. The scanning element can be designed in such a fashion that two points which must fall on the analyzed component in the image are separated by a number of points which may be called "Don't Cares", because they may or may not fall on the component. They cover a distance h. The plot of the number of events against distance h is called a covariogram. The first part of the covariogram contains information on the shape of the individual structures. The second part contains information on the distance between these structures and any periodicity in the image is revealed. The covariogram is an extremely powerful tool to characterize an image and with it the stereology of the specimen under investigation, especially if it is established in different directions.

Rose of Direction. Multi-directional image analysis by means of incremental image rotation leads to a plot called the rose of direction. It reveals any directional anisotropies as in Fig. 5.

Two-Dimensional Image Analysis. Two-dimensional analysis is accomplished by the use of two dimensional scanning elements of varying size, for example, in the shape of a hexagon. They permit processing the image so that specimen parameters which were originally inaccessible can be extracted. Two-dimensional logic is used to erode individual structures for image cleaning and to separate touching particles in order to arrive at a proper count as shown in Fig. 6. The reverse process leads to a dilation of image structures which is used to fuse conglomerates and to build bridges between particles which belong together as in the case of broken stringers in rolled steel (Fig. 7). The combination of erosion with a dilation in order to reconstitute the true area of the eroded particle is called opening illustrated in Fig. 8. This image transformation leads to an analysis of a particle population which is almost identical to a sieve analysis.

The most important advantage of television image analysis is speed. A disadvantage is limited gray level differentiation of the vidicon or plumbicon tube. There are applications which require the differentiation of many more gray levels than exist within the capabilities of a television system. In these cases, the scanning microscope photometer is employed because its photo detector, the photomultiplier tube, is in this respect far superior to the vidicon.

Scanning Photometry

A microscope photometer consists of an optical system which permits the isolation of a small specimen area for a light intensity measurement illustrated in Fig. 9.

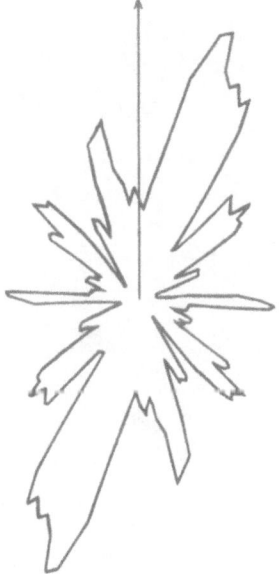

Figure 5. Rose of Direction

Figure 6. Erosion by circular scanning element

DILATION

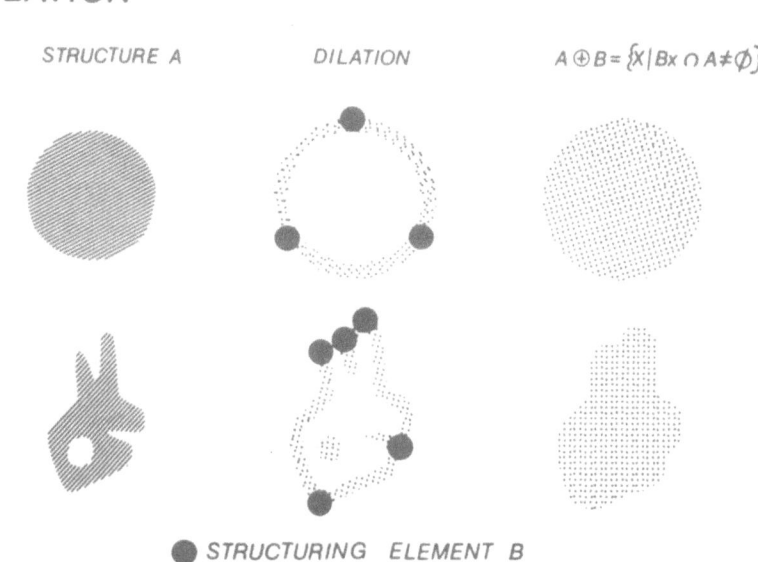

Figure 7. Dilation by circular scanning element

OPENING

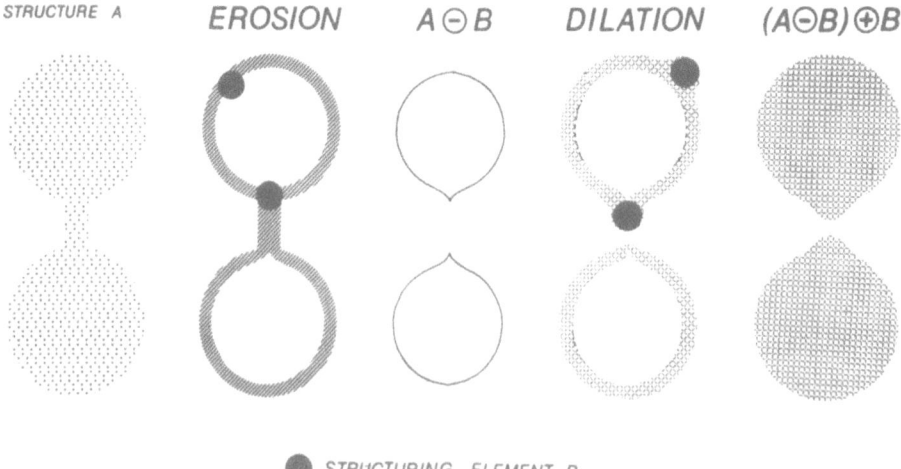

Figure 8. Opening

As a photo detector, photomultiplier tubes with different spectral responses are chosen to suit the particular application.

For scanning, step scanning stages are used. If the scanning of large areas is required, stages with step sizes of 10 micrometers and total scanning areas of up to 75 x 150 mm are available. Fine scanning stages have step sizes as small as 0.25 micrometers, but they usually cover only smaller areas. However, the combination of a fine scanning stage with a coarse scanning stage allows the fine scan of large areas. For image analysis, the scanning microscope photometer is computer interfaced and one can determine the same or even more complex specimen parameters as with the television image analyzer. It should be mentioned that a stage scanning system, as used in connection with microscope photometers, is slow as compared with a television image analyzer. On the other hand, the gray level differentiation and the capability for the analysis of individual specimen structures is superior with stage scanning systems.

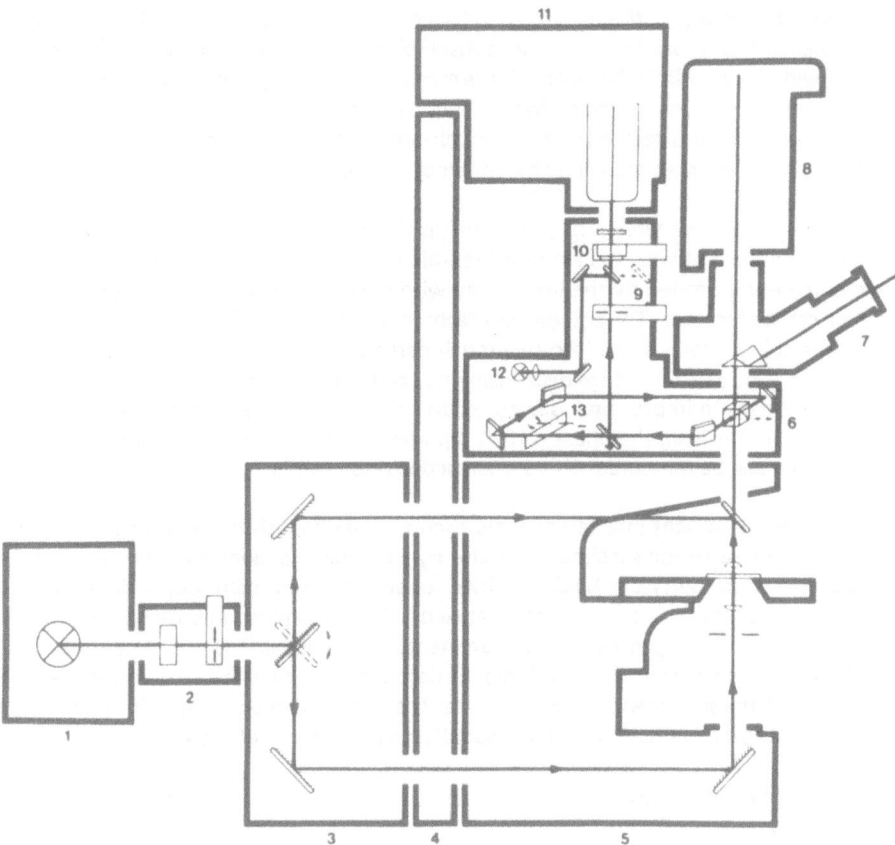

Figure 9. Microscope Photometer: 1. lamp housing, 2. lamp carrier with diaphragms and filter holder, 3. mirror housing, 4. carrying rail, 5. microscope, 6. photometry tube, 7. FSA binocular phototube, 8. camera, 9. measuring diaphragm, 10. interference filter, 11. photomultiplier with housing, 12. pilot lamp, 13. dark flap.

AXIAL MEASUREMENTS

Fine Adjustment

The use of the fine adjustment for measurements along the z-axis of the microscope is limited because, at best, one can achieve an accuracy of only ± 1 μm. It should also be realized that the fine adjustment may not be absolutely linear over its entire range. Therefore, it is desirable to calibrate the fine adjustment and to use only the linear portion for the measurement.

Interference Microscopy

Interference microscopes or attachments are used to measure much smaller distances in the Z direction, distances of a few nanometers. Their measuring range and accuracy depends on the type of interferometer and on the method used to measure the fringe displacement. Two beam interferometers permit the use of polychromatic light which means that the interference fringes have identity and, therefore, fringe displacements of more than one fringe distance can easily be recognized. The relatively wide fringe width, however, limits measuring accuracy. Multiple beam interferometers have much narrower fringes and, therefore, higher measuring accuracy. The fringe width is affected by the monochromacy of the light source and by the reflectivity of the specimen and the reference mirror.

In most interferometers, the phase shift is determined directly by measuring the fringe displacement with the aid of a filar micrometer eyepiece. This type of measurement is limited by the inaccuracy with which the center of the interference fringe can be located. The equidensity technique, a photographic procedure accomplishes a split of each fringe into two much narrower lines, the equidensites of the first order. As a result, fringe displacements can be measured much more accurately. Another means to improve measuring accuracy is the measurement in interference contrast. In this case, the phase shift is represented by a difference in image intensity, which can be measured with a microscope photometer.

In transmitted light interference microscopy, the phase shift is not only affected by the thickness of the specimen, but also by its refractive index which means that this specimen property can be determined for structures of microscopic dimension. An interesting application of the transmitted light interference microscope is the evaluation of surface profiles by measurements on a replica. The fringe contrast and the measuring accuracy in an incident light interferometer depends on the reflectivity of the specimen. If one replicates the surface and analyzes it in transmitted light, the accuracy remains constant regardless of specimen reflectivity.

Polarized Light Microscopy

Measurements with the polarizing microscope also belong into the category of axial measurements. Again phase shifts are measured as in interference microscopy but here the phase information is contained in the state of polarization of a single image point. The phase shift occurs between the two vibrations within an isotropic material. This retardation is affected by the refractive index difference of the bire-fringent material and, in transmitted light, also by the thickness. The measurement of retardation leads to the identification of a mineral or, for example, to the quantifi-cation of the strain in the glass forming a glass to metal seal. The compensators used for such measurements vary in measuring range and measuring accuracy and have cho-sen to suit the application. With Brace Koehler compensators, retardations of a few angstroms can be measured under optimal conditions, however, their total measuring range may only cover 1/30 of a wave length. Tilting compensators may have a range of 30 orders, but are far less accurate

LIGHT INTENSITY MEASUREMENTS

The basic principle of the microscope photometer was discussed earlier. Such an instrument is used for the determination of reflectance, remission, absorption or fluorescence. It frequently becomes necessary to measure, also, the spectral charac-teristics of the specimen. In this case, the photometer is combined with a mono-chromator or an interference wedge located in the illuminating beam for spectral separation and then, for example, reflection spectra can be recorded. If the deter-mination of fluorescence spectra is required, the monochromator must be inserted into the imaging beam in front of the photo detector. The applications of the mi-croscope photometer are very diversified and it is, therefore, important that the in-strument is of modular design so that it can readily be modified to suit future needs.

TEMPERATURE MEASUREMENTS

The microscope is also used to determine melting points, measure crystallization and sublimation temperatures or the phase transformation temperatures of elements or alloys.

The heating stages which are used for this purpose range from the traditional Kofler hot stage with temperature maxima of at most 400 °C to high temperature stages which elevate the temperature of the specimen to 2,500°C. The accuracy of the temperature determination depends on the temperature gradient between thermocouple and observation site. The gradient can be kept low if thermocouples with small beads are used and fused to the specimen. For temperatures above 1,800°C, it is advisable to use a micropyrometer for the measurement.

SUMMARY

In conclusion, it can be said that the microscope is an excellent measuring instru-ment. Its accuracy in some fields is indeed astonishing with lateral measurements only limited by the resolution of optics. Along the axis of the microscope, one measures in nanometer units. With the polarizing microscope, retardations in the subnanometer range can be measured. In the microscope photometer, one can measure the fluorescence emerging from just a score of molecules. The television image analyzer permits the quantitative characterization of an image and, as a result, the stereometry of the specimen. Today, the microscope has become an integral imaging element for highly sophisticated measuring machines.

ACKNOWLEDGMENTS

Illustrations by the courtesy of Ernst Leitz GmbH

REFERENCES

1. S. Flugge, *Encyclopedia of Physics,* Volume XXIX, Optical Instruments, Springer-Verlag, Berlin, Heidelberg, New York 1967.

2. Krug, Rienetz, Schulz, *Contributions to Interference Microscopy,* Hilgar & Watts, Ltd., London.

3. Rinne-Berek, *Anleitung zur allgemeinen und Polarisations-Mikroskopie der Festkorper im Durchlicht,* E. Schweizerbart'sche Verlagsbuchhandlung, Stuttgart 1973.

4. The Leitz Texture Analyzer System, Theoretical Bases and Technical Realization, *Scientific and Technical Information,* Supplement I, 4, April 1974.

DIFFERENTIAL INTERFERENCE CONTRAST MICROSCOPY

H. E. ROSENBERGER *

INTRODUCTION

The title of the microscope technique we are about to review is most descriptive and, by examining each of the words in this title, we can establish a good foundation from which to proceed into a discussion of the technical details of DIC . As it turns out, this word by word examination is best accomplished in reverse order so we shall proceed in this fashion.

MICROSCOPY

The word *microscopy* in the title is important because it is a reminder that DIC Microscopy simply is an extension of ordinary Bright Field Microscopy. To understand the theory of DIC there is no need to revise our concepts of basic microscope theory. Instead, we need only to build further onto these basic concepts. Furthermore, the basic microscope equipment required for DIC Microscopy includes all the familiar components of the ordinary bright field microscope. The only difference between Bright Field and DIC Microscopes is the modest amount of additional accessory equipment required for DIC.

CONTRAST

Probably the most important word in the title is the word *"contrast"* because it is the capacity of DIC Microscopy to induce visibly contrasting areas in the image of an otherwise featureless specimen which makes DIC such an important technique for both the biological and materials sciences. The importance of this capability arises because of the inability of the human eye to distinguish an object from its surround or background by any means other than brightness contrast or color contrast. Regardless of the differences in chemical composition between two adjacent areas of a specimen, or regardless of the height or thickness variations, if the brightness attenuation or the differential spectral attenuation of the incident light is the same for the two areas, the human eye will not detect the chemical or physical differences.

This lack of contrast in the images of specimens of known heterogeneous composition was discovered very early in the development of microscopy. In the beginning, most of the effort directed toward finding a solution to the problem was concentrated in a search

* Bausch and Lomb, Inc. Rochester, New York, USA.

for chemical dyes which would preferentially stain or etch one component of the specimen and, thus, alter its light transmissivity or reflectivity without altering that of the other components.

While chemical staining and etching of specimens to induce visible contrast is still widely practiced and is quite satisfactory for many applications, these techniques are unsatisfactory for many others. The three most frequently cited deficiencies of chemical staining and etching are (1) they are time consuming, (2) they alter the chemical and/or physical nature of the specimen and thereby are likely to introduce artifacts and (3) in the biological sciences they are unsuitable for use on living specimens. These deficiencies are serious enough to have spurred the search for "optical staining" techniques which would achieve contrast inducement through optical rather than chemical means. The search began in the last century and continues to this day. DIC Microscopy is one comparatively recent development of this continuing research into methods of contrast enhancement.

INTERFERENCE

DIC Microscopy is one of several techniques employed today which rely upon the wave nature of light to induce image contrast through the phenomenon of *interference*. Although the microscope equipments employed to execute these several interference techniques might at first glance appear to share no common relationship with each other, the fact is that the principles of all of these techniques are closely related. In each of the equipments, means are provided for dividing the source illumination and directing it along two separated coherent optical paths. Depending upon the particular equipment, the specimen may modulate the phase and amplitude of the light traveling along one of the optical paths and not the other or it may differentially modulate the light traveling along both paths. At some point subsequent to the region of the specimen the two coherent optical paths are recombined. Upon recombination, interference occurs and the details of the specimen are revealed by the nature of the resulting interference pattern.

One of the simplest interference systems we can think of is the Linnik interference microscope shown schematically in Fig. 1A. This system is designed for the examination of opaque specimens and as shown in the figure is made from an ordinary bright-field metallurgical microscope with the addition of an auxiliary objective and reference specimen. Both the auxiliary objective and reference specimen are mounted as integral parts of the microscope.

The beam divider splits each ray emanating from the illuminator into two rays. One of these two rays is reflected downward by the beam divider and then follows the usual path through the objective to the specimen where it is reflected back into the objective and thence transmitted to the eyepiece focal plane. The other ray is transmitted through the beam divider and auxiliary objective to the reference specimen where it is reflected back into the auxiliary objective and thence reflected into the eyepiece focal plane.

To see how the system works, we refer to Fig. 1B where, for simplicity, the two objectives have been deleted. A ray from the illuminator travels along the path IO to the beam divider where it is split into two rays which are directed along the two paths OA and OA'. It should be noted that the paths OA and OA' have been made exactly equal to each other so that the time required for light to travel from O to A and back to O will be exactly equal to the time required to travel from O to A' and back to O. Thus when light traveling along the two ray paths is recombined at the beam divider and directed into the eyepiece focal plane, constructive interference will occur.

If now some portion of the specimen area lies at a slightly different level from that of point A, as for example the small depression shown in the figure, the light from this area of the specimen will arrive back at the beam divider out of phase with light from the corresponding area of the reference specimen because of the longer path length it has had to travel. This light will interfere destructively and an area of reduced brightness will be formed in the eyepiece focal plane as shown in Fig. 1C.

An interference microscope designed for transparent specimens is illustrated schematically in Fig. 2. The principle of operation is the same as for the previous example, but a good deal more hardware is required. Light from the illuminator is split into two ray paths by the lower beam divider. The two ray paths then travel different but optically equal routes through a series of mirrors, condensers and objectives until they recombine at the upper beam divider. The specimen is placed in the left-hand specimen plane and a featureless reference specimen in the right-hand specimen plane. The localized minute differences in optical path in the specimen are revealed by the nature of the interference pattern they produce.

Although the word *interference* is not included as part of the title for another useful contrast inducing microscope technique, nevertheless, Phase Contrast Microscopy depends upon the phenomenon of interference. A transmitted-light phase-contrast microscope is illustrated in Fig. 3 where it is shown that the splitting takes place at the specimen itself by virtue of diffraction. The separated ray components then travel different paths through the objective. Some of these ray components are modulated both in phase and amplitude by a modulation plate built into the objective. The other components pass through the objective without modulation. The components then recombine in the eyepiece focal plane where interference occurs. Phase contrast for opaque specimens requires a substantially more complex optical system than shown in Fig. 3 and for this reason is generally available only on the large research type metallographs.

An essential requirement for achieving phase contrast is that the condenser aperture be restricted to a narrow annulus as shown in Fig. 3. This results in somewhat pronounced diffraction effects which sometimes make the interpretation of the specimen structure rather difficult. This is especially serious if the variations in optical path differences over different areas of the specimen are large. But Phase Contrast Microscopy has the advantages that the equipment is relatively easy to manufacture at modest costs, is very easy to use without the need for constant delicate adjustments and is only slightly more sensitive to ambient sources of vibration than the ordinary bright-field microscope. On the other hand, while the interference microscopes described in Figs. 1 and 2 are not subject to these same pronounced diffraction effects, they are expensive to manufacture and difficult to use. The schematic representations in Figs. 1 and 2 do not show the array of compensating adjustments required to achieve and maintain exactly equal optical path lengths between the two arms of the microscopes nor do they show the mechanical structure

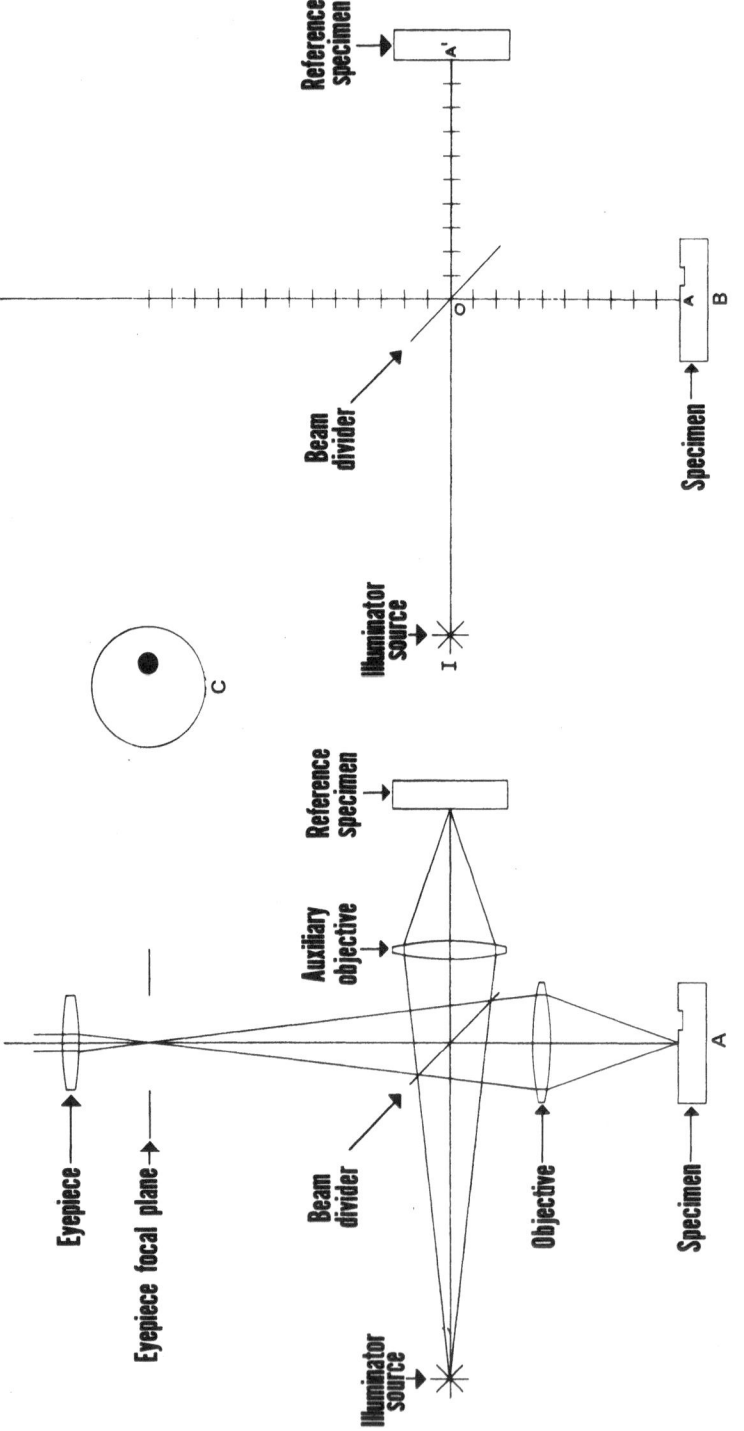

Fig. 1. Schematic of Linnik interference microscope.

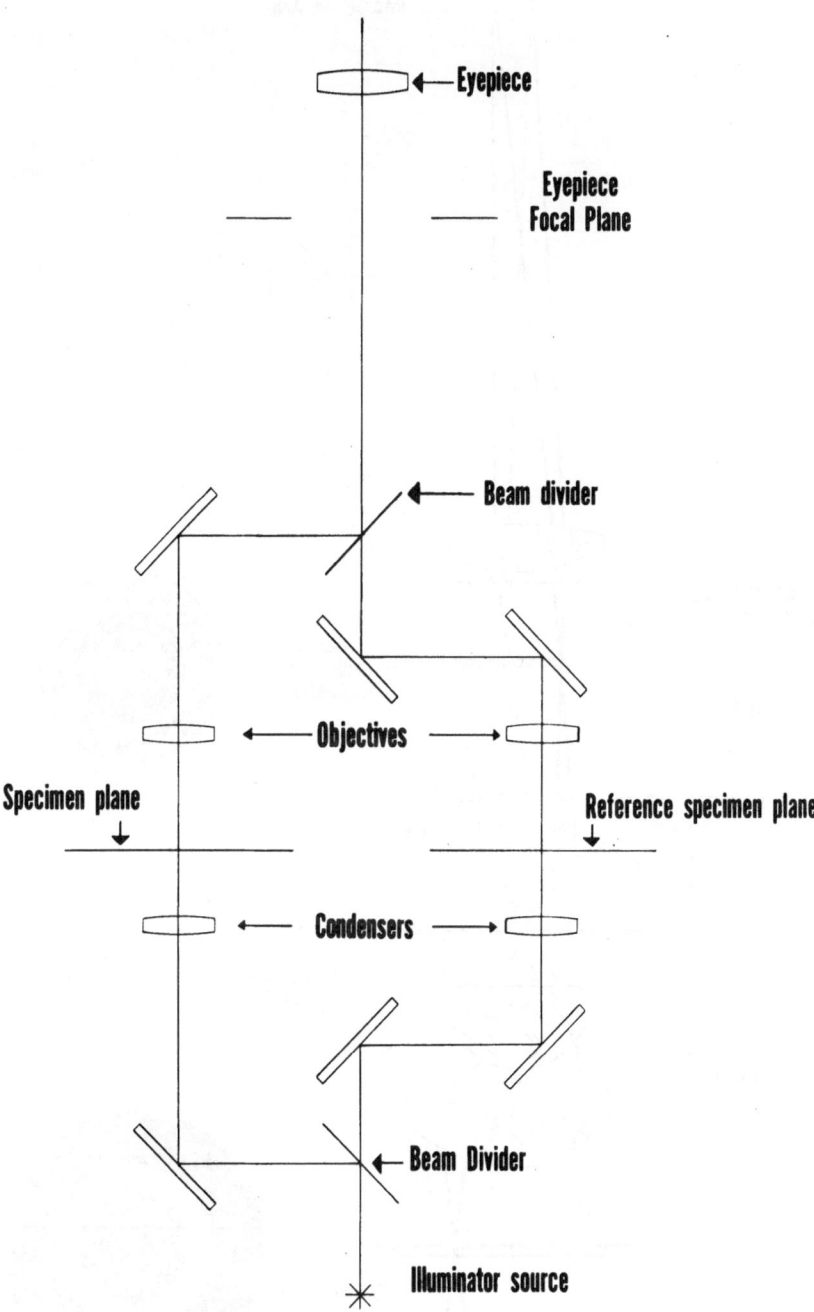

Fig. 2. Schematic of interference microscope for transparent specimens.

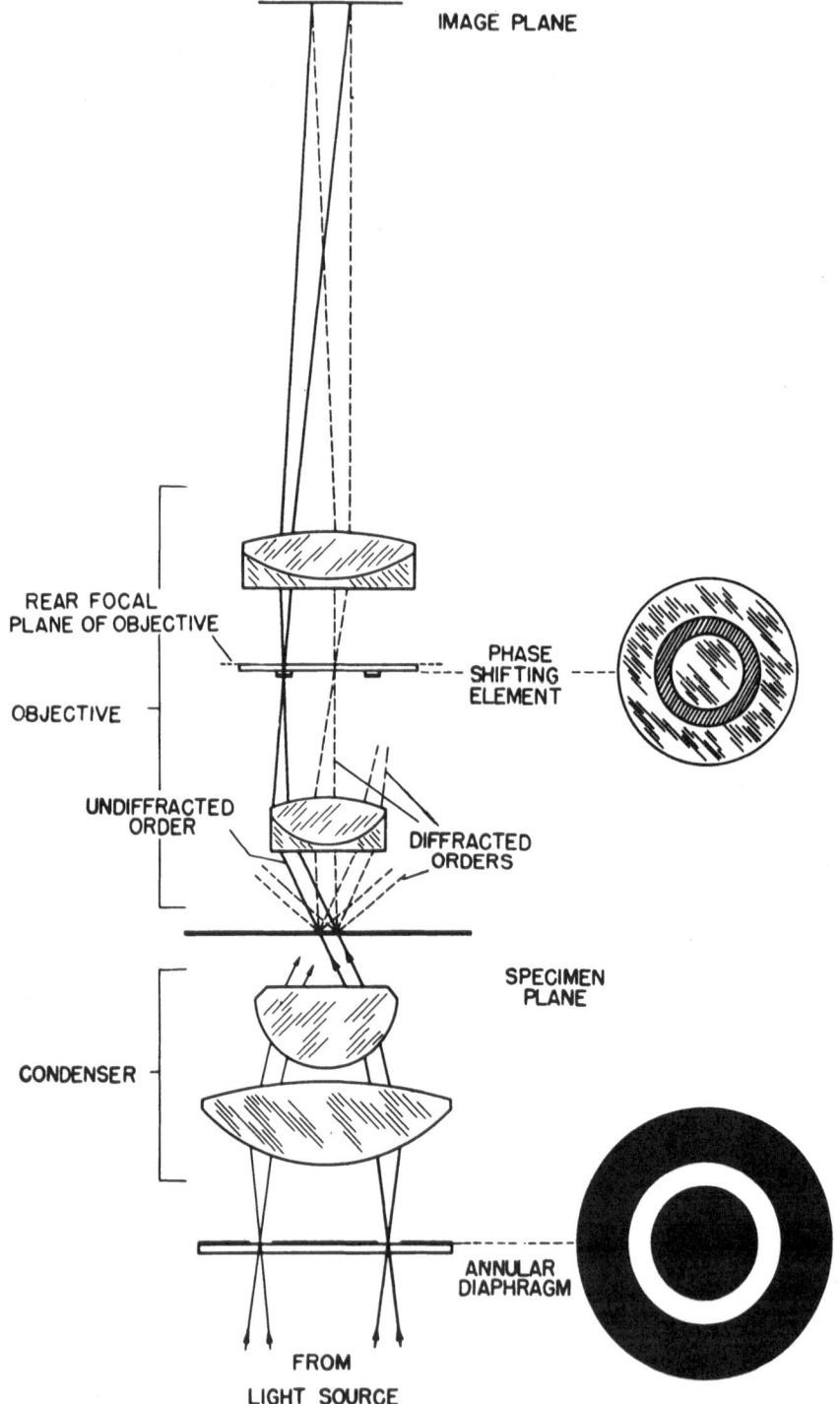

Fig. 3. Schematic of a transmitted–light phase–contrast microscope.

required to eliminate the effects of differential vibrations between the two arms. A further complication for the transmitting microscope in Fig. 2 is the requirement that the objectives and condensers must be matched for optical characteristics as nearly as modern manufacturing methods will allow. Finally, neither Fig. 1 or 2 adequately conveys the demands made on the user for the exercise of patience and skill in carrying out the necessary adjustments.

It would appear therefore that a contrast inducing microscope combining the simplicity and stability of the phase contrast microscope with the superior image quality of the interference microscope would be a valuable addition to the field of microscopy. In many ways, the DIC microscope is an embodiment of such an instrument.

DIFFERENTIAL

The word *differential* in the title is a modifier of the word *interference.* It describes the manner in which contrast is induced through interference, viz., by causing the light from each point in the specimen plane to interfere with light from a neighboring point. The manner in which this is accomplished will be described in some detail, but first, a short review of the behavior of light as it travels through several optical materials is in order.

Fig. 4A is a schematic representation of two prisms in optical contact. The material of the bottom prism has an index of refraction (index hereafter) of 1.53. If we allow the material of the upper prism to successively take on various index values from 1.00 (air) up to 1.7 or greater, we observe that a ray crossing the interface between the two prisms will be refracted to the left or to the right or not at all depending upon whether the index of the upper material is less than, greater than or equal to that of the lower prism material.

Prism combinations similar to the one shown in Fig. 4A are essential elements of the DIC apparatus except that instead of the isotropic materials used in Fig. 4A, the prisms are made from an anisotropic material such as calcite or crystal quartz.

An anisotropic material is one which exhibits physical properties which are direction dependent. Many materials exhibit this property; wood for example is easily split in one direction but resists splitting at right angles to this direction. The thermal co-efficient of expansion of some materials differs in different directions and even may be of opposite sign in different directions. Certain crystalline materials exhibit anisotropic properties in their transmission of light. In these materials there is one direction, called the optic axis, along which the velocity of light is independent of the state of polarization of the transmitted light. In all other directions the velocity of light is dependent upon the angular relationship between the optic axis and the direction of vibration of the transmitted light. Since for optical materials the index and velocity are inversely related, the index also is dependent upon this same angular relationship.

Materials which exhibit this property are said to be birefringent or double refracting because their refractive characteristics are functions of two indices instead of a single index. A prism assembly made from such a material is illustrated in Fig. 4B.

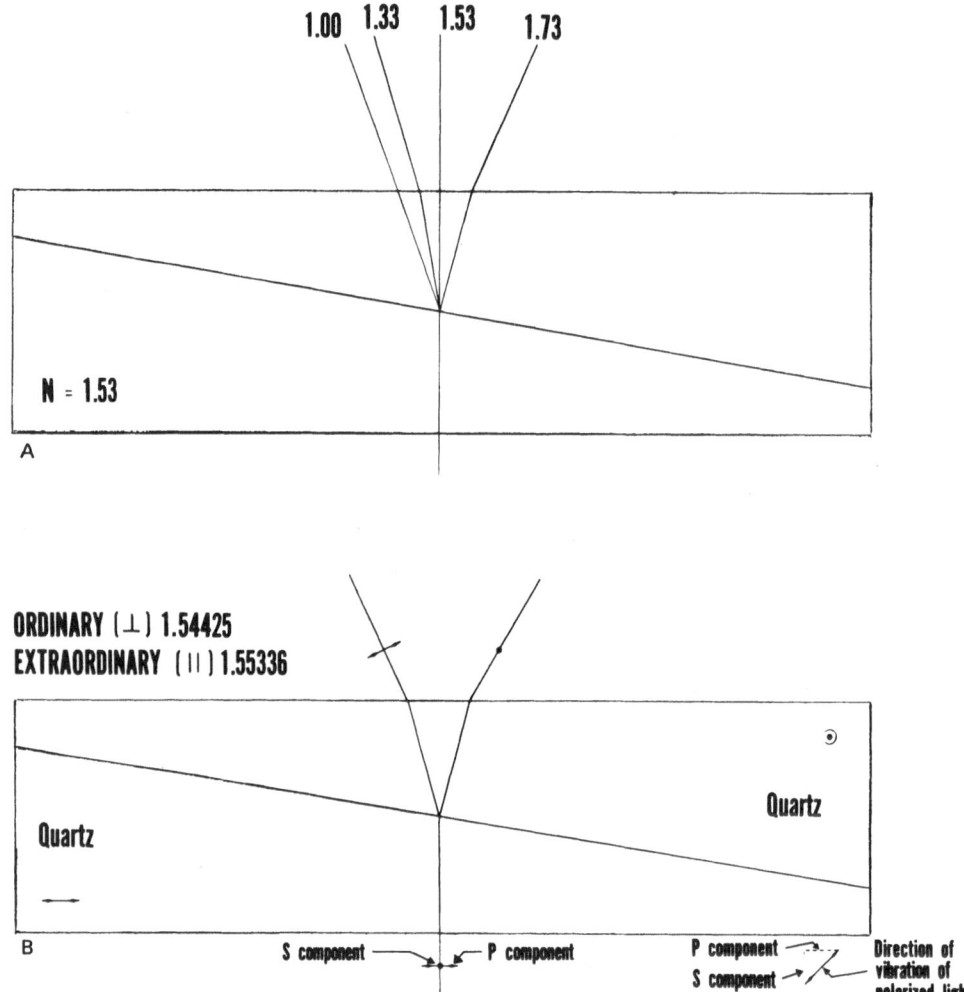

Fig. 4. Schematic of a) two prisms in optical contact and b) prisms made of birefringent material.

The lower prism in Fig. 4B is cut with its optic axis parallel to the plane of the drawing and the upper prism cut with its optic axis perpendicular. When light, plane polarized at 45° to these two optic axes enters the prism assembly, it first will be resolved into two components, one parallel (p component) and one perpendicular (s component) relative to the plane of the drawing. The index of the bottom prism for the p component is 1.55336. After crossing the interface between the two prisms the p component will be vibrating perpendicular to the optic axis of the upper prism where the index drops to 1.54425. As the p component crosses the interface from a higher to a lower index material it will be refracted to the left. For the s component the index relationship between the two prisms is reversed and therefore the s component will be refracted to the right. Thus, the single plane polarized ray which entered the prism combination leaves it as two plane polarized rays vibrating at right angles to each other. In addition to this division of a single ray into two rays, one other change takes place and this is illustrated in Fig. 5.

Referring first to the middle of Fig. 5, the p and s components leave the polarizer and reach the bottom surface of the prism assembly in phase with each other. But because of the index difference for the two components, the s component travels faster and the two components get more and more out of phase as they travel through the lower prism. As soon as they cross the interface between the two prisms, the index difference is reversed and since the physical paths through the lower and upper prisms are equal, the two components will be exactly back in phase as they leave the upper prism.

Referring now to the ray paths to the left and right in Fig. 5, the physical paths through the two prisms are unequal, hence, the two components will leave the upper prism out of phase with each other. If we increase the length of the prism assembly somewhat and add a fourth and fifth ray path, one further to the left and one to the right, we could find paths where the optical path difference for the two components would be exactly one wavelength, hence, the two components once again would leave the upper prism in phase with each other.

We refer now to Fig. 6 which is a schematic illustration of a transmitted light DIC microscope. The details of this drawing have been grossly exaggerated and some minor simplifications have been made in the ray paths for the sake of clarity.

Starting at the bottom left-hand corner of the figure, we note that entering ray path Z is split into the two ray paths labeled B and C. This prism assembly is located in the lower focal plane of the substage condenser. The two polarized components which travel along these paths, after refraction by the condenser, are directed along ray paths which are parallel to the axis of the microscope and which pass through the two separated points B and C in the specimen plane. The two polarized components are then refracted by the microscope objective to a common point in the objective exit pupil where a second prism assembly is located. The upper prism assembly is inverted relative to the lower, hence, the two polarized components are converged along the single ray path Z′ which continues on into the eyepiece focal plane.

We note then that two separated points B and C in the specimen plane are directed to a single point in the eyepiece focal plane. If now we return to the bottom of Fig. 6 and follow entering ray path Y through the system, we see that two other points C and D

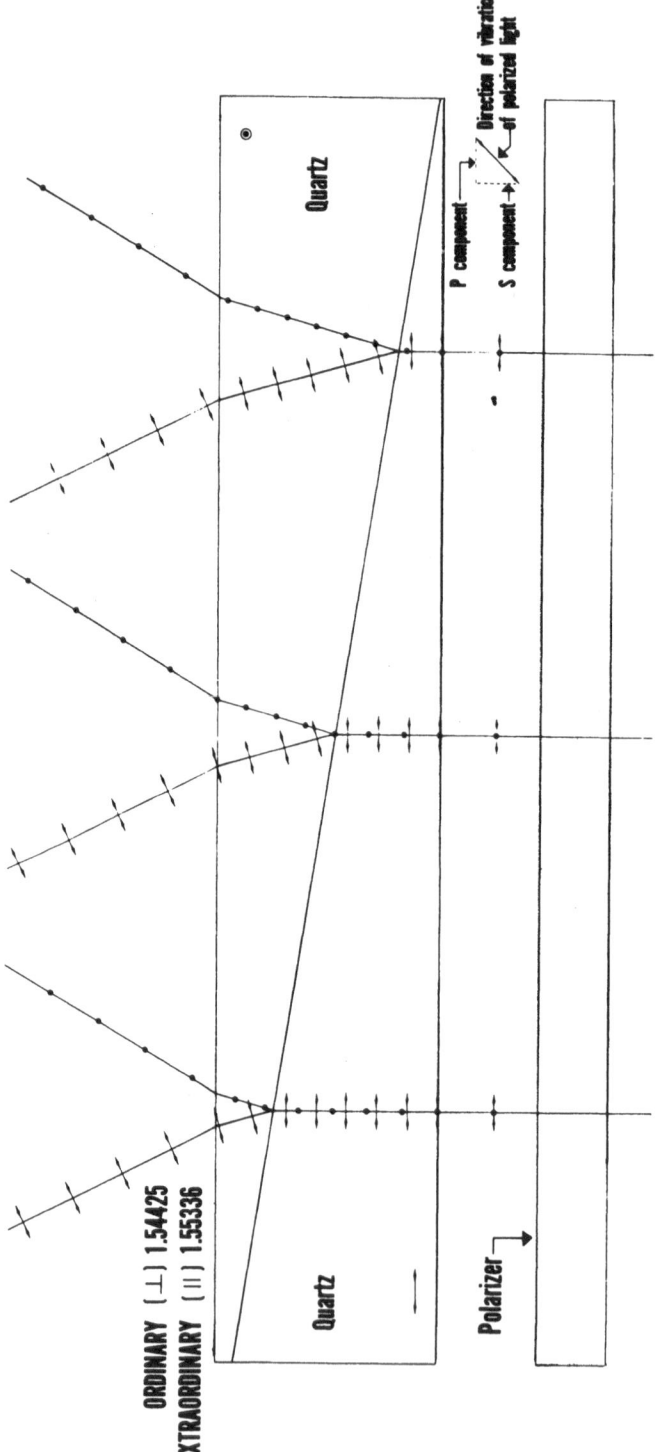

Fig. 5. Schematic of light rays through prism system shown in Fig. 4B.

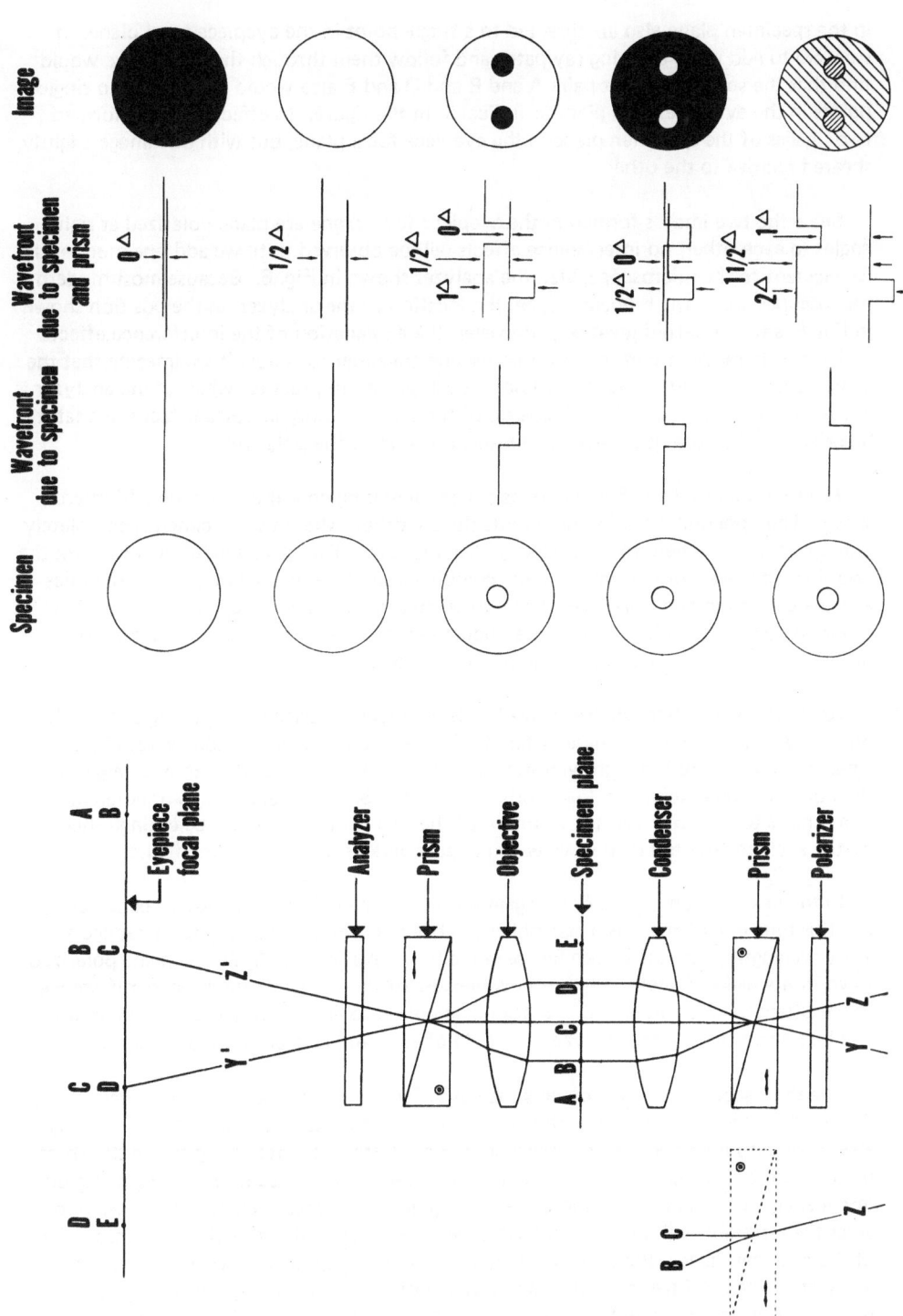

Fig. 6. Schematic of a transmitted light differential interference microscope.

in the specimen plane also are directed to a single point in the eyepiece focal plane. If we were to add more entering ray paths and follow them through the system, we would find that the specimen point pairs A and B and D and E also would be directed to single points in the eyepiece focal plane as indicated in the figure. In effect, we have formed two images of the specimen plane in the eyepiece focal plane, but with one image slightly sheared relative to the other.

Since the two images formed in the eyepiece focal plane are plane polarized at right angles to each other, no interference effects will be observed until we add one more optical element to the microscope, viz., the analyzer shown in Fig. 6. Because most modern microscopes are of the binocular type, the location of the analyzer in the position shown in Fig. 6 is the preferred location, however, the explanation of the interference effects which are characteristic of the DIC microscope are easier to follow if we imagine that the analyzer is positioned above the eyepiece focal plane. In practice, whether the analyzer is above or below the focal plane makes no difference; the only important factor is that it be oriented with its axis crossed with respect to that of the polarizer.

The right-hand side of Fig. 6 depicts several operating conditions for the DIC microscope. The uppermost section represents the condition where the specimen is completely homogeneous (or where there is no specimen present). The specimen has no effect on the two sheared wave fronts formed by the combined presence of the two prism assemblies and the condenser and objective. The two prism assemblies have been adjusted to their centered positions as illustrated in the middle section of Fig. 5, so neither prism introduces a phase difference between the two wave fronts.

As mentioned earlier, the two wave fronts are plane polarized at right angles to each other and since there is no phase difference between them they will add vectorially and emerge as plane polarized light vibrating in the same direction as when they emerged from the polarizer and entered the lower prism assembly. Since the analyzer is oriented at right angles to the polarizer, the resultant of the two components will be extinguished by the analyzer and the field of view will appear dark as indicated in the figure.

In the next section of Fig. 6, we again see a completely homogeneous specimen area, but this time one of the prism assemblies has been decentered laterally to introduce a 1/2 wavelength phase difference between the two wave fronts. Again the plane polarized wave fronts will add vectorially but now because of the 1/2 wavelength phase difference the resultant will be rotated 90°. Since plane polarized light vibrating in this direction will be transmitted by the analyzer, the field of view will appear bright as illustrated.

The third section of Fig. 6 depicts the condition where the specimen introduces a phase difference of 1/2 wavelength over a small region of its total area. One of the prism assemblies also introduces a 1/2 wavelength phase difference. Referring to the sketch of the two superimposed wave fronts we note one area where the sum of the two phase differences equals zero and one where the sum equals one full wavelength. For these two areas the resultants will be plane polarized perpendicular to the direction of transmission of the analyzer, hence, the two areas will appear dark in the field of view. For the remainder of the specimen area, the two wave fronts are 1/2 wavelength out of phase and therefore the resultant will be rotated 90° as in the previous example. This area will appear bright.

In the fourth section of Fig. 6 the specimen again introduces a phase difference of 1/2 wavelength over a small area but the prism assemblies have been readjusted to introduce zero phase difference. From an examination of the sketch showing the superimposed wave fronts, one can deduce the reason why the bright and dark areas of the field of view are reversed from those of the previous section.

In the final section of Fig. 6, the specimen again introduces a phase difference of 1/2 wavelength over a small region of its area but this time one of the prism assemblies has been shifted laterally to a point where it introduces a 1½ wavelength phase difference between the two wavefronts. When the prisms introduce a phase difference of this magnitude it is necessary to consider one other characteristic of the prism material, viz., its dispersion.

Just as the index of all ordinary optical materials varies with wavelength, so the two indices of crystalline materials also vary with wavelength. Furthermore, the difference between the two indices also varies with wavelength. Accordingly, the phase difference introduced by the prism assemblies varies with wavelength and when we say that the prism assemblies introduce a 1/2 wavelength phase difference, we really mean 1/2 wavelength difference for just one single wavelength. For all other wavelengths the phase differences will be slightly more or slightly less. When the phase difference introduced by the prism assemblies is of the order of 1/2 wavelength or less the variations in phase differences with wavelength will be very small and the effect hardly noticeable. When the phase difference introduced by the prism assemblies gets to be greater than this amount the effect is manifested by a change from a pure brightness contrast between different areas of the specimen to a color contrast. The origin of this color contrast is precisely the same as that of the color contrast characteristic of sensitive tint in Polarized Light Microscopy. It has become customary to make one of the prism assemblies long enough so that it can be made to appear in a wide range of colors.

Returning to the middle section of Fig. 6, the drawing indicates that the images of the areas where the phase difference is zero and where the phase difference is one full wavelength would appear as equally dark areas. This would be true only for monochromatic light. For white light, because of dispersion in the prism material and also because of dispersion in the specimen the left-hand dot would not appear as dark as the right-hand dot.

In our discussion up to this point, no mention was made of the exact magnitude of the shear introduced by the two prism assemblies. Fig. 6 shows a shear so large that two separate non-overlapping images of the specimen structure appear in the field of view. By showing this amount of shear in the figure the principles of DIC are more easily described and understood but in practice, microscopes with this magnitude of shear have not gained any degree of popularity.

The magnitude of the shear is determined by the two indices of the prism material and by the geometry of the prism elements. In practice the prism geometry is designed so that the point pairs AB, BC, CD and DE illustrated in the specimen plane of Fig. 6 each fall within half an Airy disc diameter of the microscope objective being used. Consequently, the two images are formed so close together that their separation is below the limit of detection. With the shear reduced to this magnitude, the image of the specimen we have been using in Fig. 6 takes on the appearance illustrated in Fig. 7. In the top section of Fig. 7 the prism assemblies have been adjusted to introduce zero phase difference between the two wavefronts. Only at the boundaries of the deformations in the two wavefronts is there a phase difference between the two wavefronts and it should be noted that this difference occurs only in the regions oriented perpendicular to the direction of shear. When the two plane polarized wavefronts are added in these regions the resultants will be either plane polarized in the direction of transmission of the analyzer or will be elliptically polarized depending upon the exact value of the phase difference. In either case the resultant will contain a component which will be transmitted by the analyzer. The specimen structure boundaries which lie in a direction perpendicular to the direction of shear will be outlined by a bright band and thus made visible against a dark background. The width of the bright area shown in Fig. 7 is greatly exaggerated.

The lower section of Fig. 7 shows what happens when one of the prism assemblies is adjusted to introduce a phase difference. An examination of the sketch of the superimposed wave fronts will show why there will be three brightness levels in the field of view. Experience has shown that this state of adjustment generally reveals the most information about the specimen structure and this image with its bright band on one side and shadow on the opposite side has come to be recognized as a kind of hallmark for DIC. This shadow effect also is responsible for the three-dimensional appearance of the image. The observer's everyday experience leads him to interpret an area having a dark shadow on one side and a bright region on the opposite as a height variation in the topography.

FINAL COMMENTS

Two final comments must be made about the equipment required for DIC. With reference to Fig. 6, it was pointed out that the upper prism assembly must be located in the plane of the objective exit pupil. In practice, this is not possible with most objectives because the plane of the exit pupil for most objectives lies below the shoulder and inside the objective tube. Moreover, the plane of the exit pupil frequently lies inside one of the objective lens elements. This is an advantageous condition for phase contrast but prior to Normarski's work it limited the application of DIC to use with a few low power objectives which had accessible exit pupils. Nomarski overcame the problem by altering the design of the prism assembly. By tilting the optic axis of one of the prism elements the ray paths through the prism assembly are modified to those illustrated in Fig. 8 where it will be noted that the point of convergence (or divergence) of the ray paths lies outside the prism assembly. With this arrangement, the prism assembly can be positioned above the objective with the point of convergence located in the plane of the objective exit pupil.

The final comment concerns the equipment required for applying DIC to the examination of opaque specimens. As shown in Fig. 9, the ordinary metallurgical microscope can be converted to DIC by the addition of a polarizer and analyzer and a single prism assembly; the latter performing the functions of both prism assemblies required in the transmitted light microscope. The same theory of operation applies for the incident light microscope, but because of the double passage of light through a portion of the incident light microscope, the illustrations become somewhat cluttered and a little more difficult to follow.

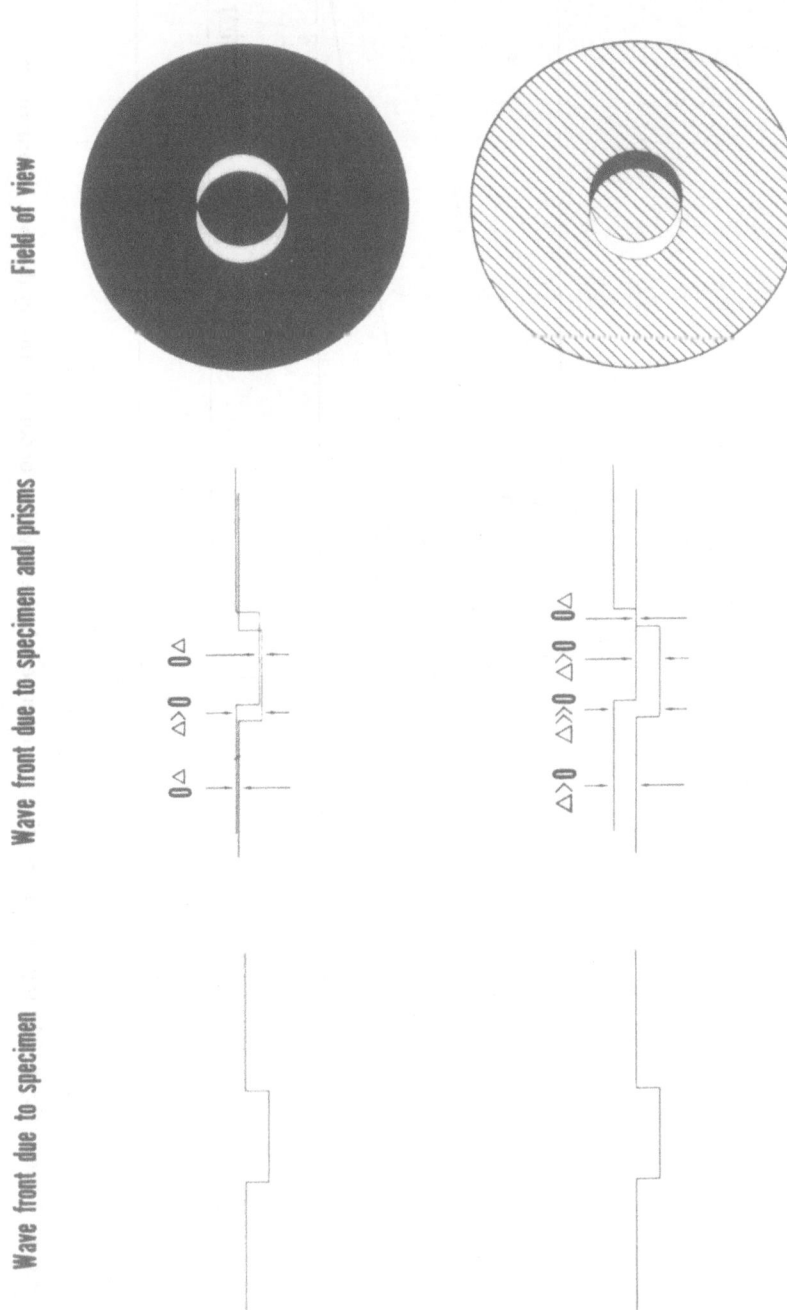

Fig. 7. Schematic of image shear that occurs in differential interference microscopy.

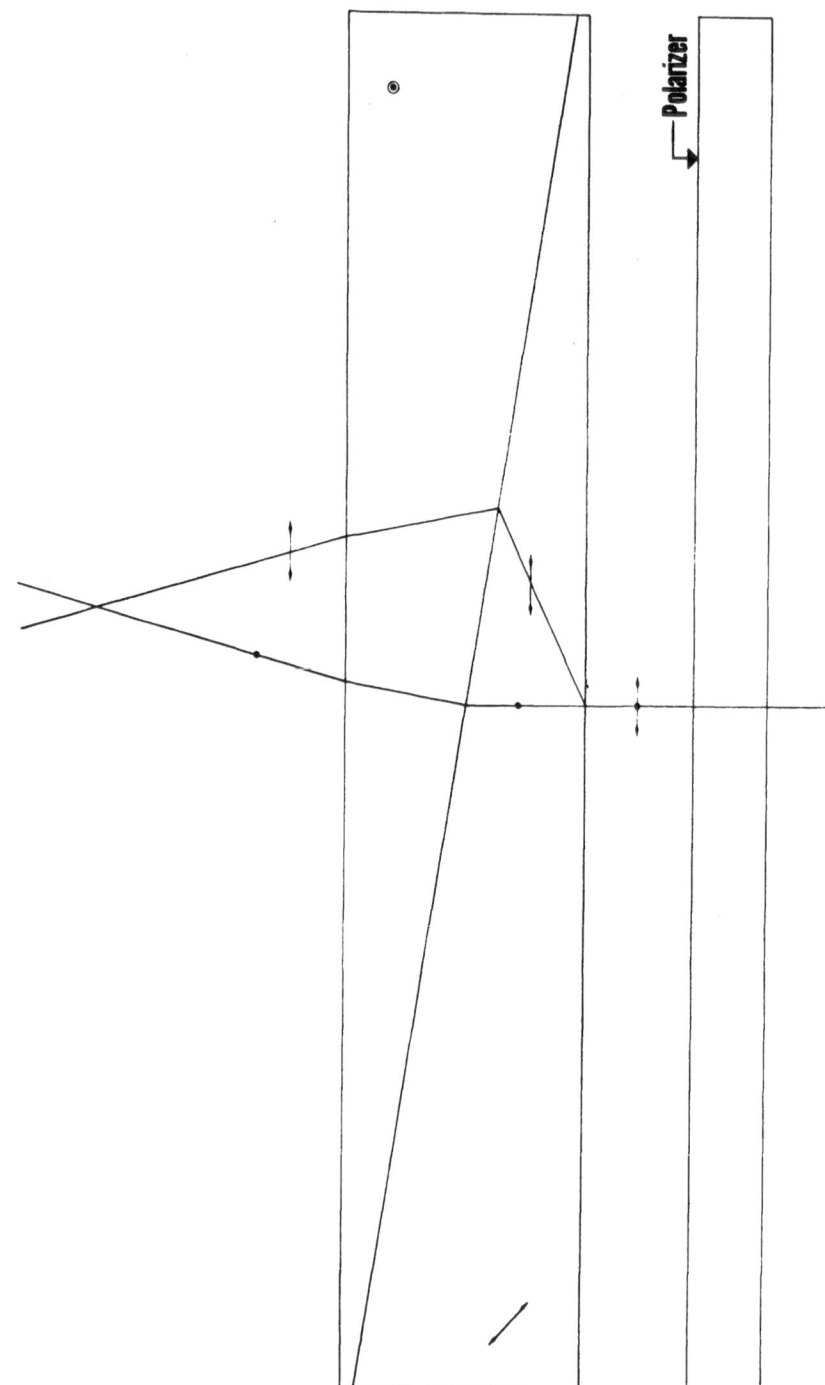

Fig. 8. Schematic of light rays in differential interference microscopy that occurs by tilting the axis of one prism.

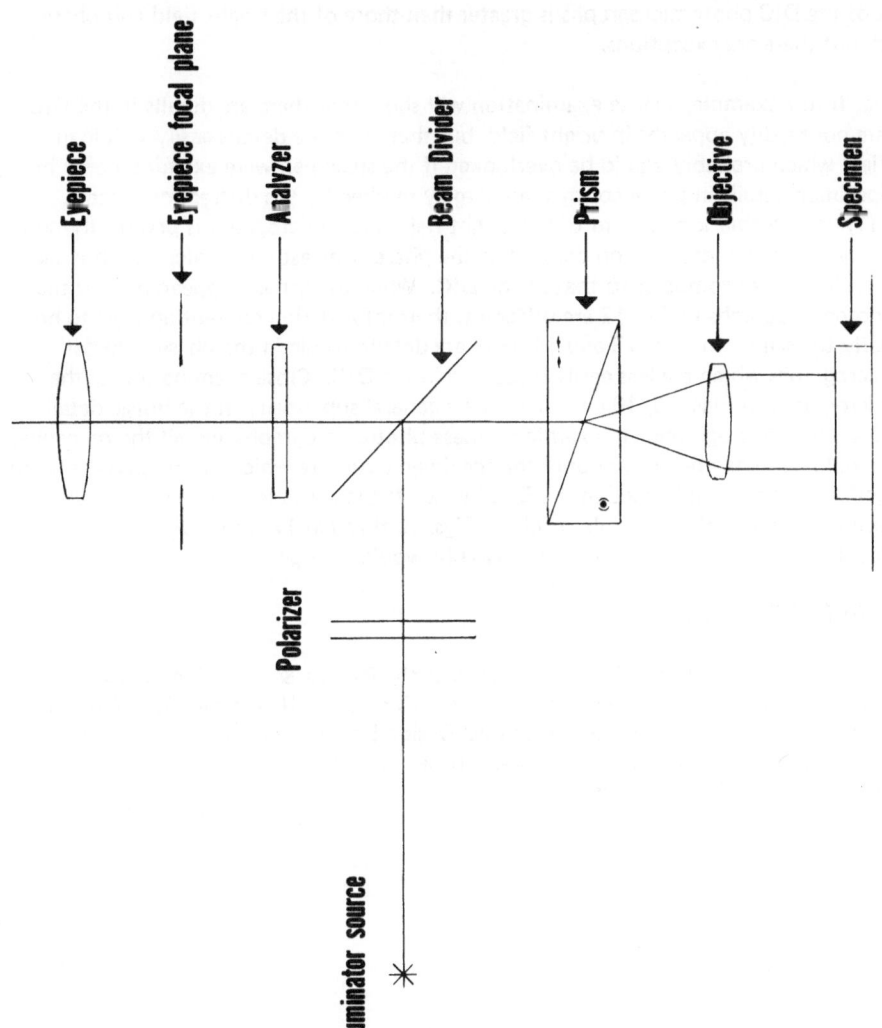

Fig. 9. Schematic of differential interference microscope for examining opaque specimens.

APPLICATIONS

Figs. 10 through 17 are a series of photomicrographs in which a comparison can be made between the images of various materials as they appear in bright field, phase contrast and DIC. The direction of shear in the DIC photomicrographs is approximately from the lower left-hand corner to the upper right. Generally speaking the information content of the DIC photomicrographs is greater than those of the bright field and phase contrast, but there are exceptions.

In Fig. 10 for example, a close examination will show that there are details in the DIC which are not readily apparent in bright field, but there also are details easily visible in bright field which probably would be overlooked if the specimen were examined only in DIC. Specimen details in phase contrast are largely masked by the diffraction effects. In Fig. 11, the information content of the bright field photomicrograph is practically non-existent and while the information content of the phase contrast photomicrograph is significant, it in no way compares to that of the DIC. While the general appearances of the three photomicrographs in Fig. 12 are different, their information content appears to be very nearly the same. In Figs. 13 and 14 there are details visible in the phase contrast photomicrographs which are less easily discerned in the DIC. Close examination of the photomicrographs in Figs. 15, 16 and 17 shows a general superiority in the image detail of the DIC photomicrographs. In examining these photomicrographs and all the preceding ones it should be remembered that only the specimen structure which lies perpendicular to the direction of shear is enhanced in the DIC image. If the specimens were rotated 90° and reexamined, some of the details visible in Figs. 10 through 17 would disappear while other details not visible in these photomicrographs would emerge.

ACKNOWLEDGEMENTS

The author wishes to thank Mr. B.N. Iannone of the Bausch & Lomb Metallurgical Laboratories who prepared the specimens for Figs. 10 through 16 and Mr. D.E. Judd of the Bausch & Lomb Optical and Electro-optical Design Department who provided the photomicrographs. The equipment used was a Research II Metallograph with integral phase contrast and DIC capabilities.

FERRITE

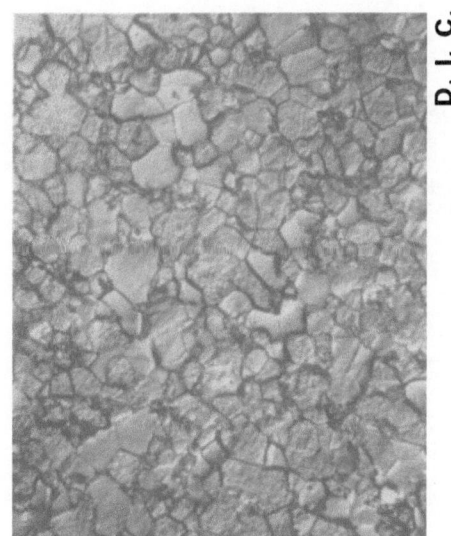

D. I. C.

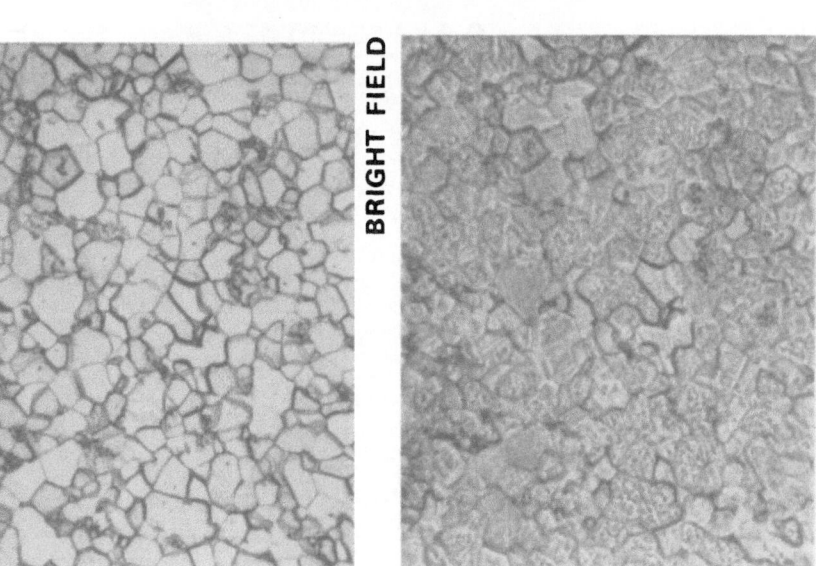

BRIGHT FIELD

PHASE CONTRAST

Fig. 10. Photomicrographs of ferrite at 200X. Specimen was mechanically polished and deep etched.

CAST ALLOY

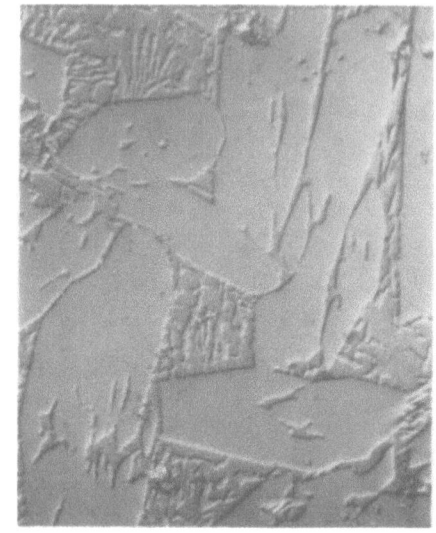

D. I. C.

BRIGHT FIELD

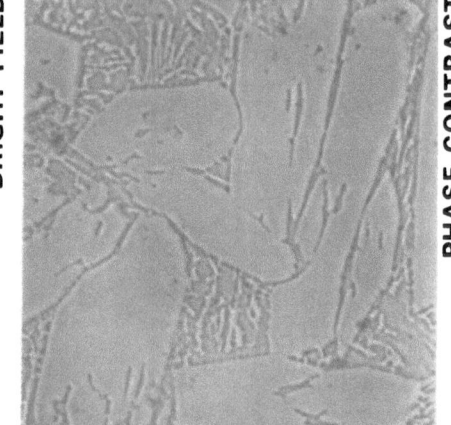

PHASE CONTRAST

Fig. 11. Photomicrograph of cast alloy at 200X. Specimen was mechanically polished, unetched.

Fig. 12. Photomicrograph of cadmium at 100X. Specimen was chemically polished.

HIGH MANGANESE CAST STEEL

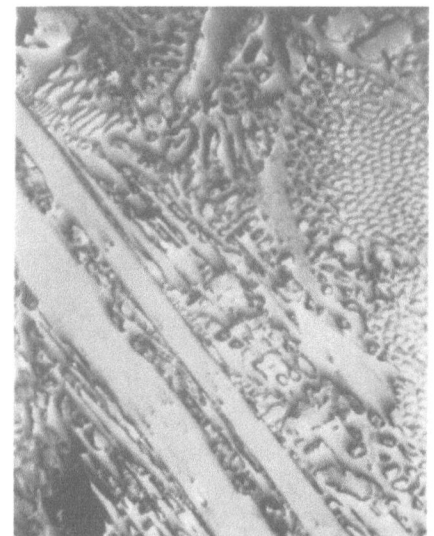

D. I. C.

BRIGHT FIELD

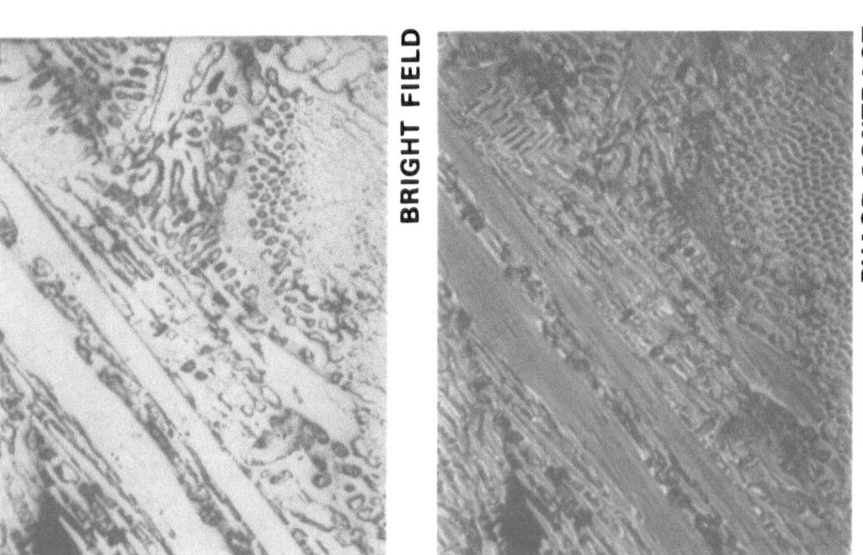

PHASE CONTRAST

Fig. 13.   Photomicrograph of high manganese cast steel at 125X.   Specimen was mechanically polished, unetched.

BETA BRASS

D. I. C.

BRIGHT FIELD

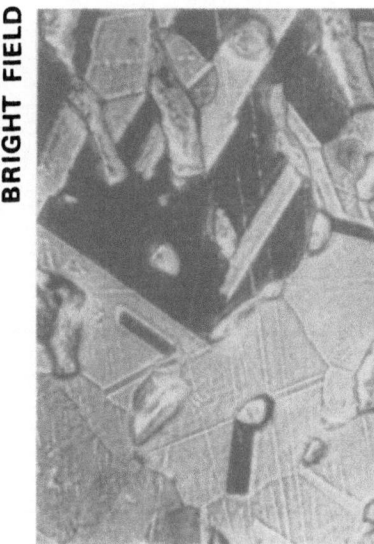

PHASE CONTRAST

Fig. 14. Photomicrograph of beta brass at 400X. Specimen was mechanically polished and etched.

ZIRCALOY

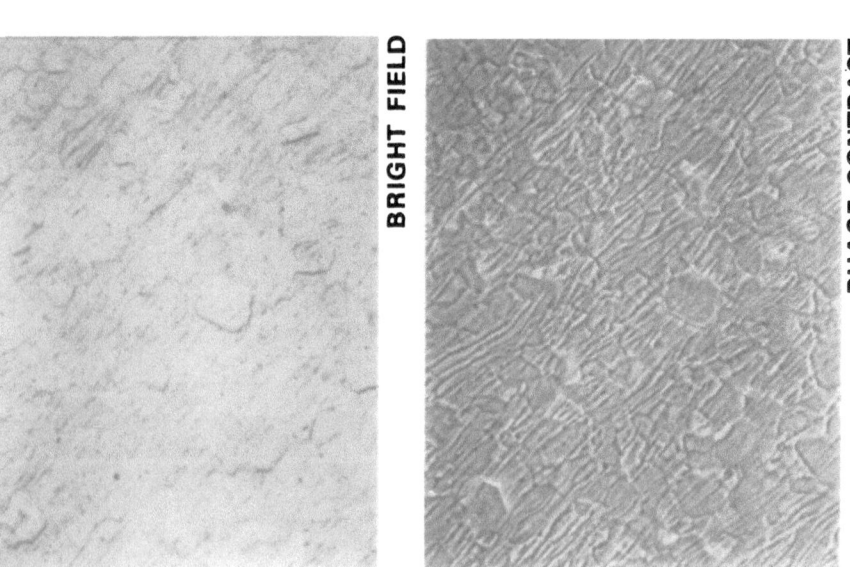

Fig. 15. Photomicrograph of zircaloy at 400X. Specimen was chemically polished.

ZIRCONIUM

D. I. C.

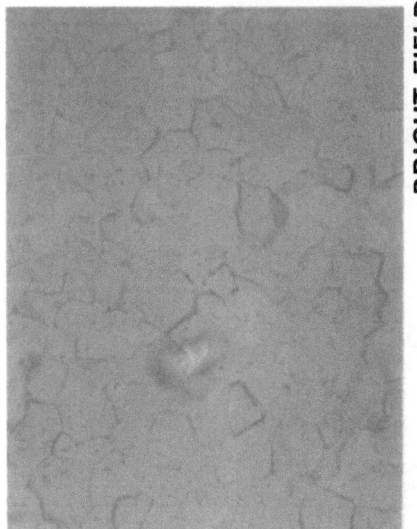

BRIGHT FIELD

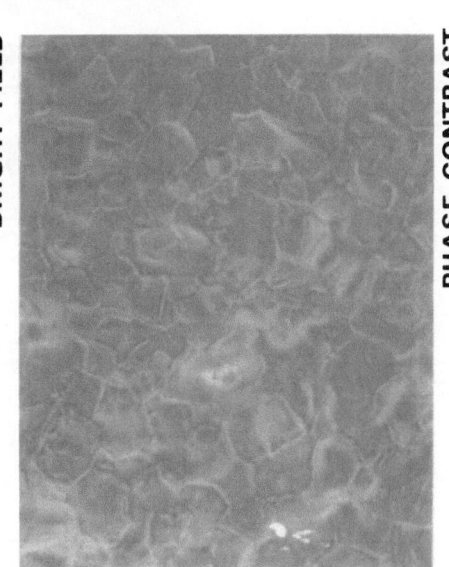

PHASE CONTRAST

Fig. 16. Photomicrograph of zirconium at 400X. Specimen was chemically polished.

INTEGRATED CIRCUIT

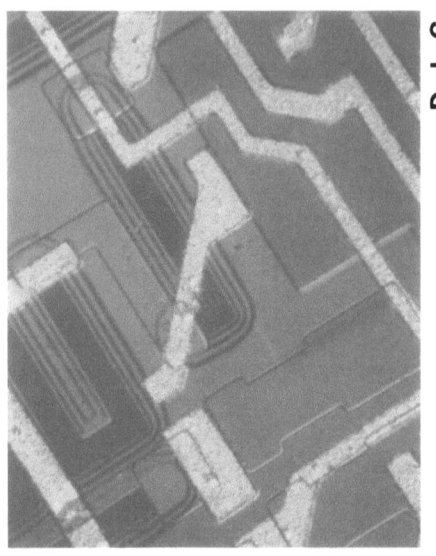

D. I. C.

BRIGHT FIELD

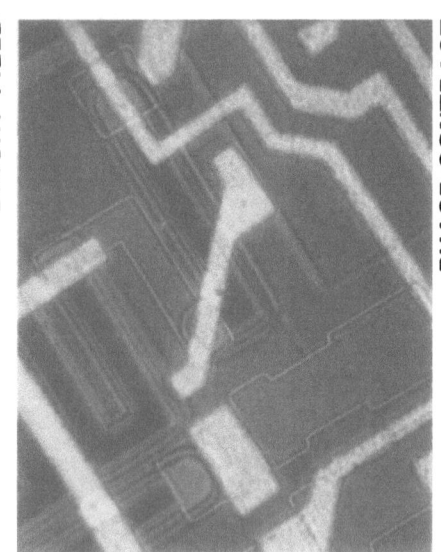

PHASE CONTRAST

Fig. 17. Photomicrograph of integrated circuit at 200X.

SURFACE TOPOGRAPHIC CHARACTERIZATION EMPLOYING OPTICAL METHODS;

SURVEY OF QUANTITATIVE AND QUALITATIVE METHODS

H. E. KELLER *

Determination of the location of an infinite number of points on a surface in a 3-coordinate system results in a description of its complete topography.

There are basically four quantitative optical techniques to determine topography, namely —

Focusing technique: Any microscope with a graduated focusing drive and a graduated mechanical stage can be used for this method. The differences of vertical adjustments for a sufficient number of points will give us the topography. This measurement depends:

    a. on the accuracy and linearity of the focusing drive

    b. on the depth of field of the microscope system. This prime influence contains a physiological, subjective factor, the visual acuity and can therefore only be approximated. For a circle of confusion to appear under a viewing angle of two minutes, the depth of field is given by the following formula:

$$D = \frac{1000}{7M(NA)} + \frac{\lambda}{2(NA)^2}$$

where:
D = Depth of field in μm
M = Total magnification of microscope
NA = Numerical Aperture
λ = Wavelength of light used

Examples: For objective 40/0.85 at 800x Magnification = 0.58 μm
For objective 100/1.3 at 800x Magnification = 0.3 μm

This technique is of course slow and tedious but can be accurate to within 0.5 μm. The range depends on the focusing travel of the microscope. A disadvantage is of course the small field which can be viewed only, since objectives with high numerical aperture and magnification are used. (See Fig. 1).

Stereoscopic technique: In this technique essentially the same principle is used as that employed for evaluation of aerial photographs (Stereophotogrammetry).

A Stereomicroscope is equipped with two precisely aligned "contact marks" in the intermediate image plane of both eyepieces. When viewing through the microscope, these marks appear at the intersection of the axes of the two Stereolightpaths. They are brought into optical contact with the surface point investigated (See Fig. 2).

* Carl Zeiss, Inc., New York , New York 10018 USA.

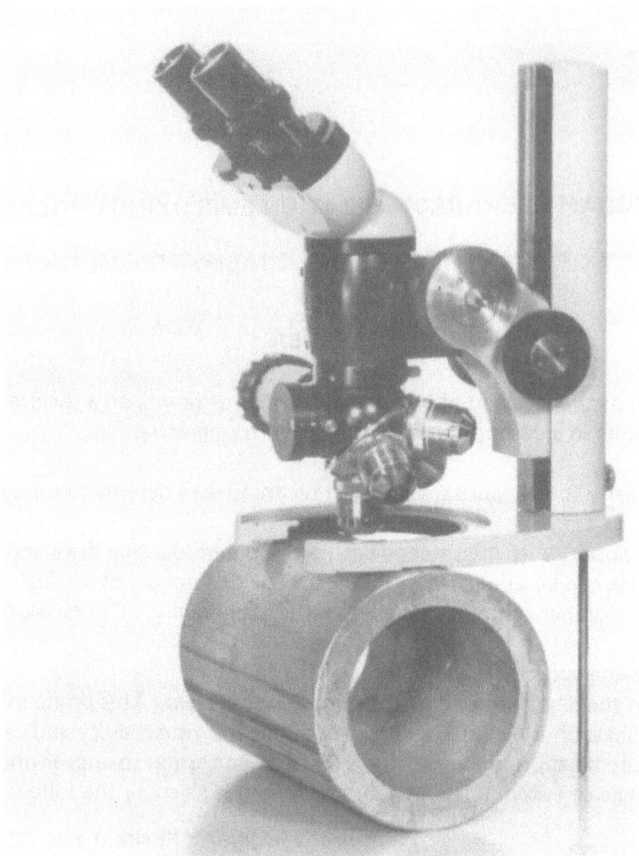

Fig. 1. Special depth measuring microscope with graduated focusing drum.

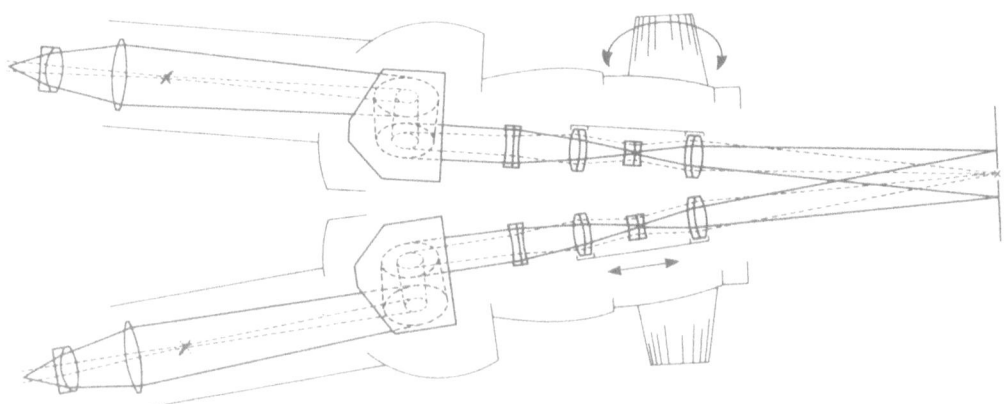

Fig. 2. Light path of a Zoom-Stereomicroscope. Contact marks indicated by "X".

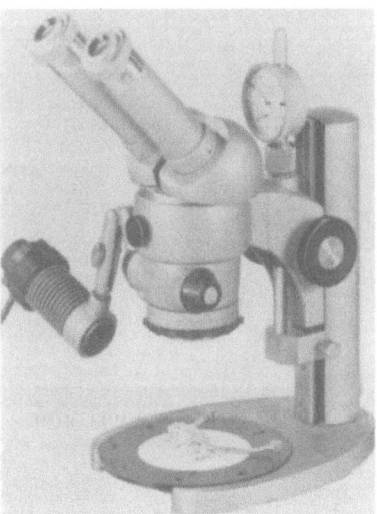

Fig. 3. Stereo-microscope on special stand with dial gage for depth measurements.

Again, either a graduated focusing movement or a dial gage attached to a special stand measures the vertical displacement as optical contact is established with different surface points. (See Fig. 3).

Lines of equal elevation can be traced by using an x − y stage connected to a plotter.

The advantages of this technique are obvious:
>A large field can be viewed at low magnifications and good depth of field obtained, yet measurements can still be carried out with good accuracies.

The accuracy depends again to a large measure on the operator's experience and on the stereo-angle employed.

Like the focusing technique, the stereoscopic method as a point by point technique is slow and cumbersome.

Light Section Technique: A slit is projected at a $45°$ angle of incidence unto the surface under investigation. It is observed at an angle of $90°$ to the plane of incidence. (See Fig. 4).

The bright line viewed on the surface exactly traces the surface profile along the line's intersection with the surface. Upon sectioning at $45°$, this profile appears enlarged by a factor of $\sqrt{2}$ and can be measured with an eyepiece micrometer or recorded on film.

Measurement with this system is far more objective and consequently more repeatable than those previously mentioned. Accuracies and measuring ranges depend on the magnification and the numerical aperture of the objective used and on the field of view of the system.

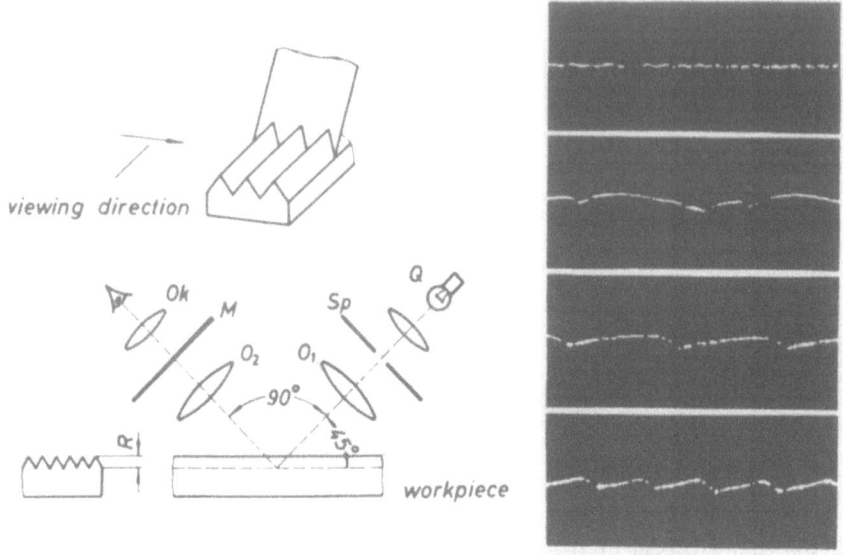

Fig. 4. Principle of light section technique and 4 examples of different surface profiles.

Accuracies to ±0.25 μm and a measuring range from 1 to 400 μm can be obtained. (See Figs. 5, 6 and 7).

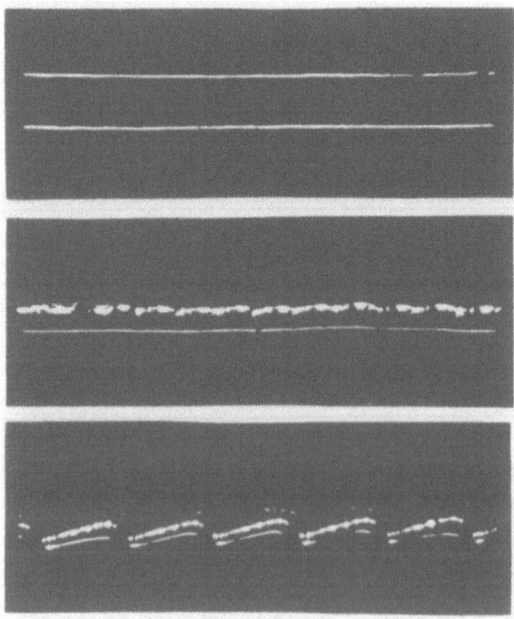

Fig. 5 . Examples of light section images of transparent coatings.

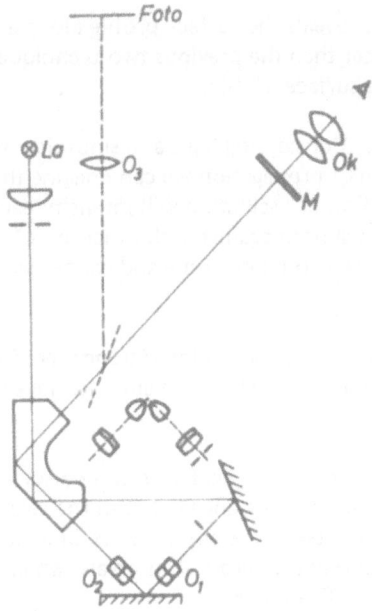

Fig. 6. Light path through a light section microscope.

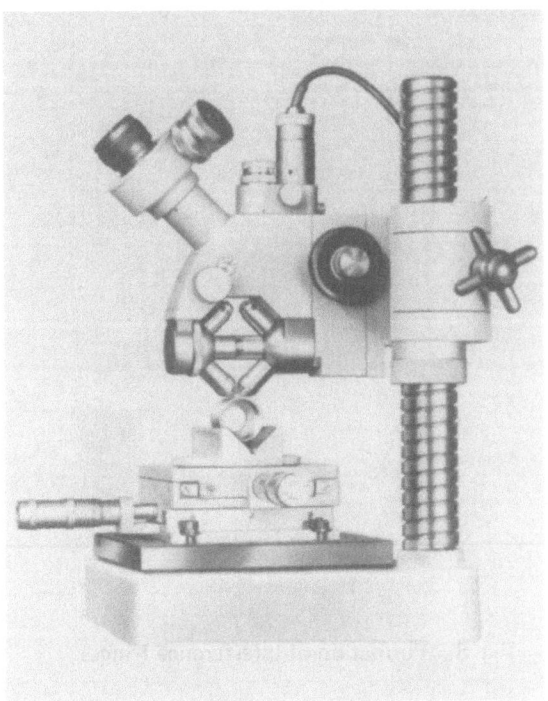

Fig. 7. Light section microscope

The light section technique reveals the surface profile along a cross-section through the surface, is considerable faster than the previous two techniques and allows for an instant photographic record of the surface profile.

Interferometric Technique: Considering light as a sinusoidal wave motion oscillating at right angles to the direction of propagation we can imagine that two superimposed waves in the same state and direction of oscillation will intensify each other (constructive interference) while two waves shifted against each other by $\lambda/2$ (half the wavelength) will vibrate in opposite directions at the same time and cancel each other out, so called destructive interference.

To obtain total constructive or destructive interference requires light of the same wavelength and light which is coherent — where the emitted wave packages are in phase, or light from a single source.

Now it can be stated that wavefronts reflected from two surfaces (reference and sample) will cancel each other wherever the distance between the two surfaces is exactly a path difference of $\lambda/2$ between the two wavefronts. Since in incident light one beam traverses this distance twice, darkness appears at intervals, which correspond to distance or elevation differences of $\lambda/2$. (See Fig. 8).

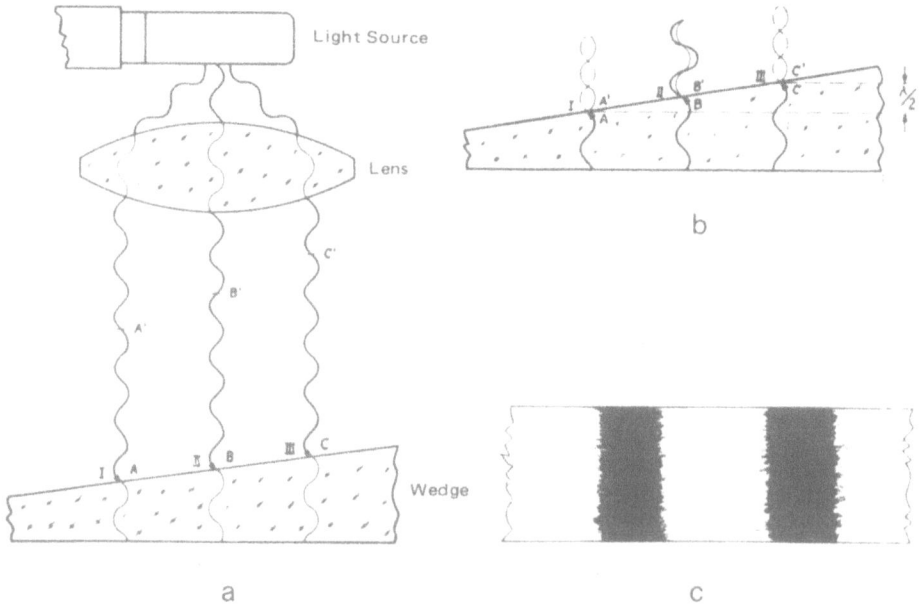

Fig. 8 . Formation of interference fringes

Interference fringes (dark lines of destructive interference) appear along areas where the elevation difference between the sample and the reference is $\lambda/2$ or multiples thereof.

The best analogy is to compare the interference fringes to lines of equal elevation on the topographic map. In this instance, sea level is the reference plane while in interferometry a mirror or beam divider, depending on the type of interferometer, acts as a reference. (See Figs. 9 and 10.)

Fig. 9. Topographic map and interference fringe pattern.

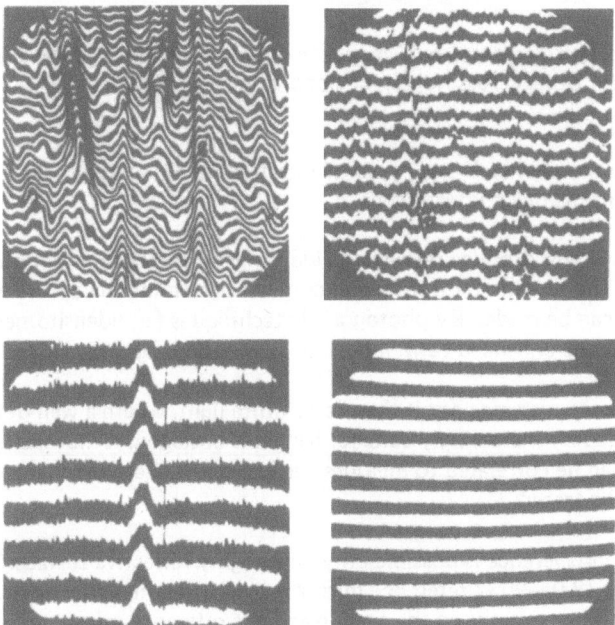

Fig. 10. Typical double beam fringes — photographically enhanced.

Changing positions of the reference mirror can of course result in a completely different fringe pattern for the same surface. (See Fig. 11).

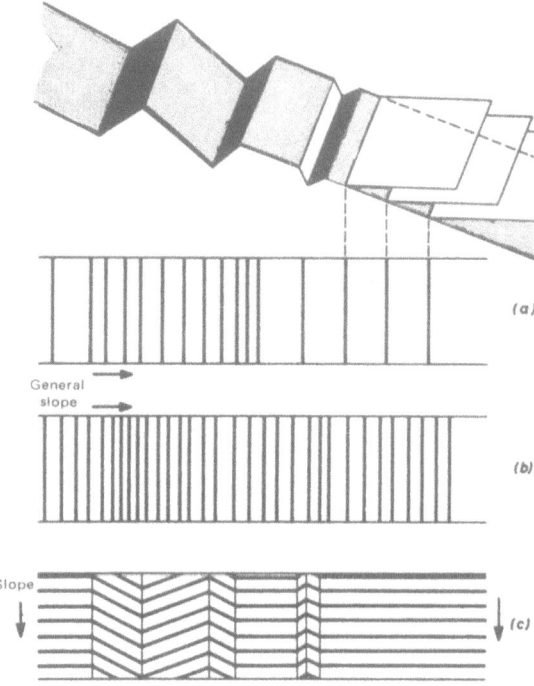

Fig. 11. Change of fringe pattern with change of slope between reference and surface
a and b: slope across ridges; c: slope parallel to features

We can differentiate between two types of interference in incident light: double beam— and multiple beam interference.

Double beam systems produce relatively wide fringes directly superimposed over surface features. Their width, definition and contrast determine the accuracy with which measurements can be made. By photographic techniques (equidensitometry) double beam fringes can be enhanced to well defined lines. (See Figs. 12, 13 and 14).

Good double beam systems permit the use of white light, where a well-defined black fringe surrounded by increasingly colorful fringes is obtained. This results in "coded" fringes which can be correlated to measure vertical steps where monochromatic fringes cannot be traced.

Multiple beam interference is possible only with highly reflective surfaces. An equally highly reflective (8 0 —90%) reference mirror is placed on the surface. Multiple reflections of wavefronts between the sample and the reference mirror result in a signal

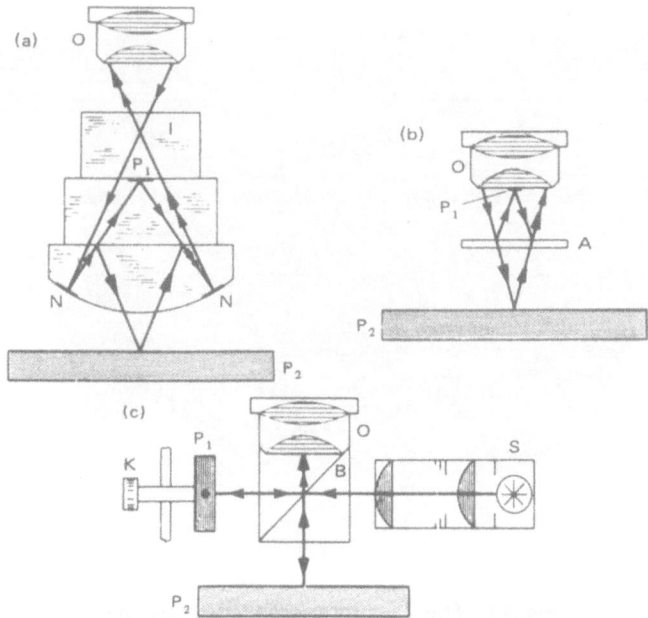

Fig. 12. 3 types of double beam interferometers:
a: Dyson   b: Mirau   c: Watson (Michelson)

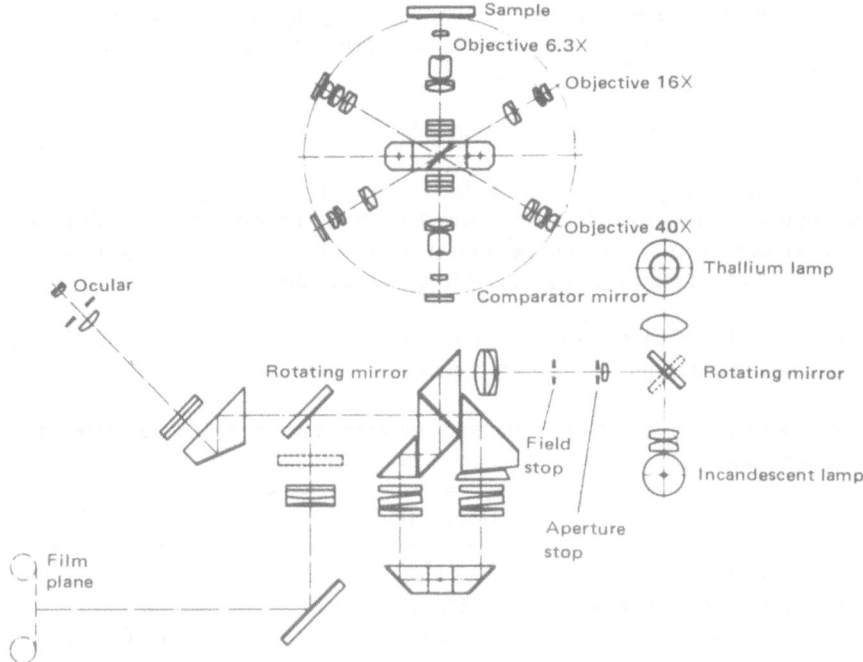

Fig. 13. Light path of Michelson Linnik type Interference microscope.

Fig. 14. The Zeiss Interference Microscope

amplification so that the darkest portion of the interference becomes a thin, well-defined hairline.

Now fringe distances and fractions thereof can be measured very accurately. Due to the multiple reflection there is, however, a slight shift between the fringes and their region of origin!

Fig. 15 shows a typical multiple beam interference pattern.

This system can be relatively easily added to existing incident light microscopes. Recommended are special L.D. objectives designed and corrected to view through the reference mirror. The reference mirror is mounted either to the objective or to the stage in an arrangement that allows it to be focusable and tiltable.

Mounting to the stage means less chance to damage the surface and permits the combination with other microscopic techniques.

Accuracies obtainable with the different interferometer techniques are given in the following table:

Evaluation by	Double Beam	Multiple Beam
Visually direct	$0.10 \lambda/2$	$0.1 \quad \lambda/2$
Fixed eyepiece micrometer	$0.05 \lambda/2$	$0.02 \quad \lambda/2$
Filar micrometer	$0.02 \lambda/2$	$0.005 \lambda/2$
Equidensitometry	$0.01 \lambda/2$	$0.005 \lambda/2$

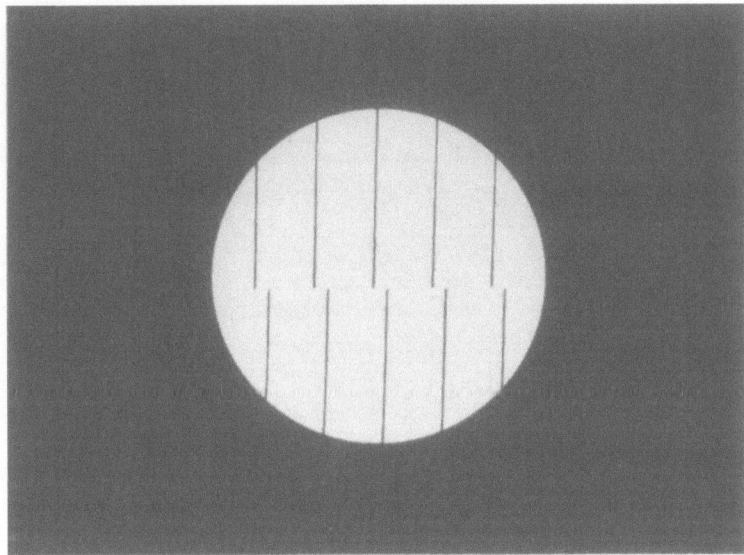

Fig. 15. Multiple beam pattern across a sharp step

Of the quantitative optical techniques, the light section— and interferometric methods complement each other in range and accuracy. Both are essentially noncontact and non-destructive and, as such, ideal for the measurement of soft and sensitive surfaces.

Qualitative techniques for the detection of surface topography are being discussed in other papers. For the sake of completeness, brief mention should be made of some of the most commonly used methods:

Oblique illumination: Shifting the illumination aperture off center can give a good shadowing effect. It may, however, cause uneven illumination and result in loss of resolution.

Modulation contrast: A surface gradient detection system whereby gradients in opposing directions are assigned substantially different amplitudes or intensities.

Differential interference contrast: Also a very sensitive surface gradient (elevation gradient) detection system using interference between two slightly sheared (separated) wave fronts. This technique permits continuously adjustable contrast.

Both modulation contrast and differential interference contrast render images much like oblique illumination but with greater sensitivity and without the loss of resolution.

Double beam interference: A flat plane specimen can be made to appear dark or light by spreading the interference fringes on any double beam system infinitely far apart by setting the reference mirror parallel to the surface investigated. In mono-chromatic light features which deviate from the perfect flatness will appear in differ-ent gray tones while in white light they will appear in different colors.

MICROHARDNESS TESTING AND HARDNESS NUMBERS

ANTON F. MOHRNHEIM *

INTRODUCTION

In this paper, the Vickers and Knoop hardness test methods for the determination of microhardness are discussed and a simplified method for comparison of numerical hardness number values are presented. Microhardness testing enables the testing of thinner and harder materials than ever possible by macrohardness testing. Increased demands on the strength of miniature parts, and a growing interest in the initiation and growth of microcracks of materials, as well as required close quality control of protective surface coatings, have created a great need and interest for miniload hardness testing.

MICROHARDNESS TEST APPLICATIONS

In metallographic work, the hardness of minute areas of precision parts, fine wire, thin metal and foil, and for the characterization of individual particles of inclusion in metallic microstructures is of great importance. Microhardness testing, in particular by the Knoop method, has expanded to materials other than metals. New areas have been opened up for research applications in new materials: jewels (both natural and synthetic), glasses, ceramics, enamels, paints, and plastics. Microhardness tests have become a simple, fairly accurate, and relatively inexpensive research tool in materials research.

BRIEF HISTORICAL REVIEW OF MICROHARDNESS TESTING

While the Vickers method originated about 1925 in England, a more sensitive method was suggested by F. Knoop [1] of the U.S. Bureau of Standards in 1939. A commercial instrument was first marketed about 1950 and is still extensively used in the U.S. under the name ' Tukon ', a trade name of the manufacturer [2]. A Standard Method of Test for Vickers Hardness of Metallic Material [3] was adopted in 1952 by ASTM [4].

About another twenty years passed before the Knoop microhardness test gained its present recognition. For instance, a well-reputed Handbook of Microscopy [5], published in 1968 , did not mention the Knoop hardness test method, nor did the name appear in the 1968 Index of ASTM. However, about this time, standard Knoop methods [6,7] of test for microhardness of materials were being discussed and adopted. An exam-

---
* University of Rhode Island, Kingston, Rhode Island.

ination of published literature using the Lockheed Dialogue Retrieval Service [8] with Compendex (COMPuterized ENgineering InDEX) found that since 1972 about fifty-two papers on Vickers microhardness and eight papers on Knoop microhardness testing [9-16] have been abstracted. It may be inferred from the names of the authors, the paper titles, and some additional informative keywords on the paper content on the latest references on Knoop microhardness testing that an internationally wide interest exists in Knoop microhardness testing and that it has been used for a large number of applications. For detailed information, the abstracts of the Engineering Index should be read or the original papers consulted.

THE VICKERS MICROHARDNESS TEST

The Vickers microhardness test uses a calibrated machine to force its indenter, which is a square-based pyramidal diamond with face angles of 136 degrees (± 30 min), into the surface of the material under test; the length of the two diagonal indentations are then measured optically. Vickers hardness number, VHN, is computed from the following equation [5] —

$$VHN = P/A_s = 2 P \sin (a/2)/ d^2 \qquad \text{(Eq. 1)}$$

where P = load in kilogram — force, kg—f
           A_s = surface area of indentation, mm^2
           d = diagonal length of indentation, mm
           a = face angle of indenter, $a/2$ = 136 deg

Since the units presently used in microhardness testing are gram—force, g—f, and micrometer (1 μm = 0.001 mm) rather than kg—f and mm, Equation 1 is more conveniently expressed as —

$$VHN = 1854.4 P_1/d_1^2 \qquad \text{(Eq. 2)}$$

where P_1 = load in gram—force, g—f
      d_1 = average diagonal length of one indentation, μm.

Though Equation 2 uses metric units in the traditional sense, the force and pressure units, as adopted by the international system of units (SI), requires familiarity with the unit measures of force in newtons (N) and of pressure in pascals (Pa); that is, 1kg—f = 9.80665 N, exactly, and megapascal, 1MPa = 1MN/m^2. (The SI symbol "M" stands for the prefix mega, in this context indicating that the unit area 1 m^2 = 1,000, 000 mm^2.

THE KNOOP MICROHARDNESS TEST

The Knoop microhardness test uses the same, or a similar machine, to force its diamond indenter into the surface to be tested. The indenter is ground to a pyramidal form in which the longitudinal angle is 172½ deg and the transverse angle is 130 deg. The length of the long diagonal is measured optically in making a measurement; this length being about 7 times the length of the short diagonal. The Knoop hardness number, KHN, is obtained by dividing the applied load, in kg—f, by the projected area of the indentation, in mm^2, computed from the measurement of the long diagonal of the indentation. Conveniently, the following equation is used —

$$KHN = 14229 \ P/L^2 \qquad (Eq. 3)$$

where 14229 .... a constant based on the National Bureau of Standards specifications for Knoop indenters.

P .... indenting load in g—f

L .... length of the long diagonal in μm.

CHARACTERISTICS OF VICKERS AND KNOOP HARDNESS

At equal load, the Vickers diamond penetrates about 4 to 5 times as deep into the surface to be tested than the Knoop diamond. It has been discussed [17] in detail that if surface layers of equal thickness are to be investigated, equal penetration depths should be strived for by selecting a load for the Knoop tests which is 2.39 times the Vickers load. For example, HV 200 (Vickers hardness using a load of 200 g—f) should be compared with HK 500 (Knoop hardness using a load of 500 g—f) in order to arrive at approximately comparable penetration depths (Table 1). If however, thin surface layers with different hardnesses and different thicknesses are involved, generally the Knoop method is preferred; for instance, the hardnesses of surface zones formed during case hardening, nitriding and other surface treatments. The quality of the surface on which the hardness is to be measured has to be considered when making microhardness measurements. In general the measurement of the long diagonal length of a Knoop impression can be made more accurately than measurement of the length of the Vickers diagonals, and therefore, the Knoop is generally less variable with regard to surface quality.

TABLE 1. Comparisons of Vickers and Knoop

Feature	Vickers	Knoop
Factor for computing hardness number	1854 (Eq. 2)	14229 (Eq. 3)
Recommended g—f range, approx.	200—15000	25—1000
Depth of indentation, t in μm	$d_1/7$	L / 30
g—f ratio to obtain the same t $\approx$ 2/5, e.g.	200 g—f	500 g—f

COMPARISON OF MICROHARDNESS NUMBERS AND ADDITIONAL CONSIDERATIONS

Vickers and Knoop hardness numbers are calculated by using Equations 2 and 3, respectively, or are found directly in Standard Hardness Conversion Tables [18]. However, VHN and KHN cannot be fitted to a single conversion relationship for all metals. Indentation hardness is not a single fundamental property, but a combination of properties, and the contribution of each to the hardness number varies with the type of test. The modulus of elasticity has been shown to influence conversion at high hardness levels; at low hardness levels conversions between hardness scales, measuring depth and the measured diameters are likewise influenced by differences in the modulus of elasticity. Separate conversion tables for steel, high nickel alloys and brass are required. Table IV of "Hardness Conversion Numbers for Nickel and High-Nickel Alloys" [18] shows the ratio KHN/VHN to be very close to 1.14 for a large range of indenter loads. Figs. 1 and 2 of this reference show plotted versus the larger corresponding KHN for nickel and high-nickel alloys.

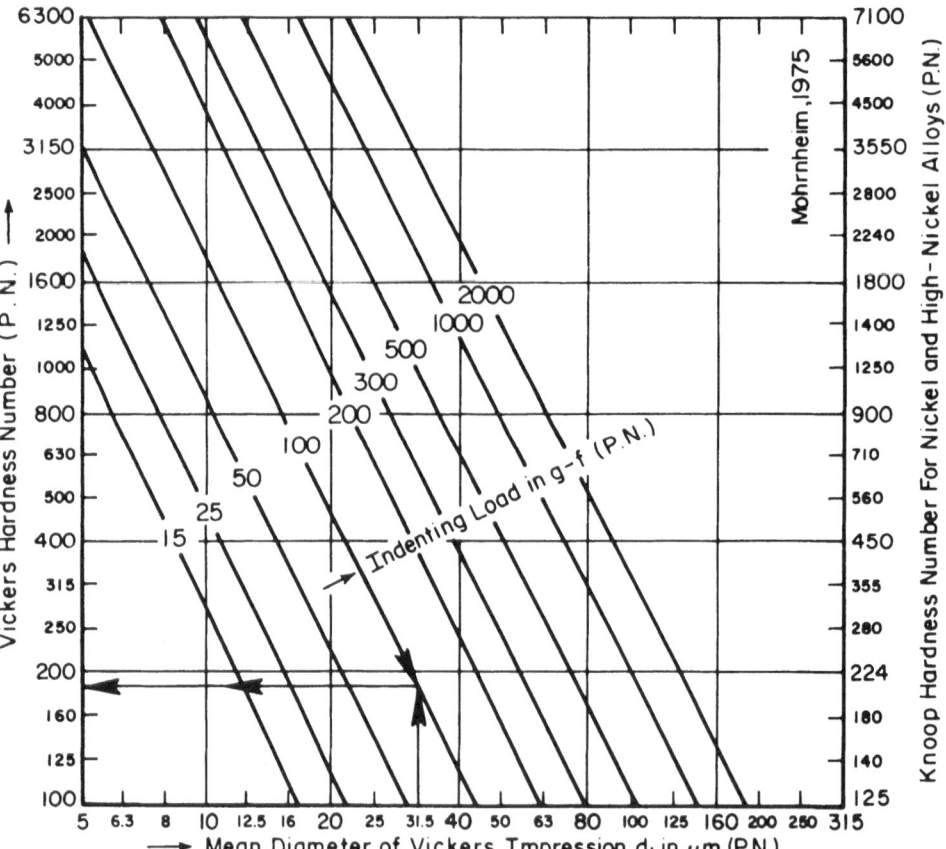

Fig. 1. Graphical approximation of Vickers hardness number (left ordinate) from the measured diameter of the Vickers diamond impression (abscissa) and the indenting load (diagonal lines of equivalence). Simplified log plot featuring preferred numbers (P.N.). For solution see text.

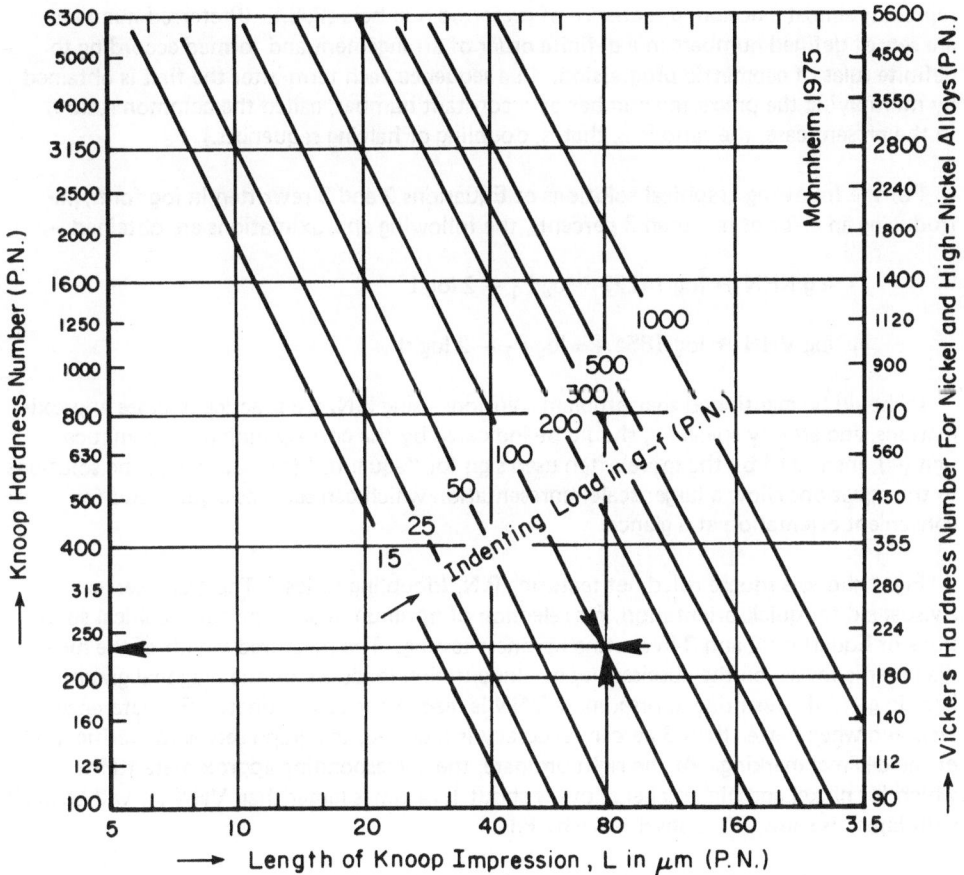

Fig. 2. Graphical approximation of Knoop hardness number (left ordinate) from the measured length of the long diagonal impression (abscissa) and the indenting load (See Fig. 1).

Graphical conversions are difficult since the equations involved in the conversion are power functions producing curved lines on a uniform scale plot. They become, however, straight lines when log—log scales are used [19]. If the evenly divided scales show instead of the log value the corresponding mantissa value (antilog) then very simple straight line plots can be drawn on ordinary graphs having square grids as will be shown and explained for the microhardness numbers [20].

The loads normally supplied with the Miniload Hardness Tester* are 25, 50, 100, 200, and 300 g—f. Test loads of 500, 1000 and 2000 g—f are available upon request. Consequently, microhardness tests may be carried out under loads from 25 to 2000 g—f, and

* Manufactured by E. Leitz, Inc., Germany

represent almost a doubling sequence of preferred numbers (P.N.). (Preferred numbers are sets of defined numbers in a definite order of arrangement and formed according to definite rules of geometric progression. In a sequence each term after the first is obtained by multiplying the preceding number by a constant number, called the common ratio r. In the present case, the ratio is 2, that is, doubling or halving sequences.)

For the following graphical solutions of Equations 2 and 3 rewritten in log form, (introducing an error of less than 3 percent), the following approximations are obtained —

$$\log KHN \approx \log 14229 + \log P_1 - 2 \log L$$

$$\log VHN \approx \log 1854.4 + \log P_1 - 2 \log d$$

It should be mentioned that problem solutions using P.N. are practically close approximations, and strictly speaking, should be indicated by the corresponding mathematical sign ($\approx$), instead of by the more often used sign for "equal to" (=). Graphs of the solutions to the equations allow a larger scale representation which can serve as a guide and for convenient orientation at a glance.

Fig. 1 shows a square ruled net featuring P.N. (doubling series). The graph serves as a visual aid for quick orientation, for selection of optimum loads, and for graphical solutions of Equations 2 and 3 in routine hardness testing. The measured length of the mean diameter is shown on the abscissa in μm. Indentation loads are on the diagonal graph lines in g—f; the resulting approximate VHN is read at the left ordinate. For reference tests and when better than 3 percent accuracy is required, the graph serves for verification of the decimal marking. At the right ordinate, the corresponding approximate KHN, which for nickel and high-nickel alloys is about 1.14 times larger than VHN (as will be dealt with later), is shown for convenience by P.N.

In order to find an unknown VHN using the graph, start with the determined mean length of the two indentation diagonals, e.g., d = 31.5 μm on the abscissa. Then move vertically up to the diagonal graph line that corresponds to the applied load, e.g., P = 100 g—f. Then, at the height of the intercept read the lefthand scale, e.g., between 160 and 200 VHN by following the heavily drawn directional arrows. Therefore, for the example the result computed from Eq. 1 is that VHN = 186.9. The difference may be inconsequential for most technical applications.

Analogous to Fig. 1, Fig. 2 serves for the graphical solution of Equation 2. The heavily drawn directional arrows show an example; for L = 80 μm and P = 100 g—f. The left most arrow points to VHN between 200 and 250 g—f/mm^2, that is, the preferred number 224 vs. the computed value 220.

CONVERSION OF MICROHARDNESS TO MACROHARDNESS

Microhardness numbers as obtained by the method of Vickers and Knoop cannot be directly compared with the macro hardness numbers as obtained on the traditional macrohardness scale of Rockwell C. It has been earlier pointed out that this was a basic international problem of materials testing [21]. Much work has been devoted in order to establish curves for the conversion from Knoop hardness to Rockwell C hardness and vice-versa. The results were of limited value since the significance and correlation of the variables were not, and still are not, fully understood. The influencing parameters as mentioned before in part are varying indentation depths, work hardening of the surface during sample preparation, grain size and strength [22], vibration of the instrument during indentation, visibility errors, and elastic recovery of the indented material at the ends of the long diagonal. The duration of the period of contact (dwell time) of the load on the surface has been found to be insignificant [17], as was the subjective random error of the measurements of the diagonal length by different observers [13]. Homogeneity of the calibrated hardness test blocks, however, is of importance and must be ascertained.

In 1969, Batchelder [17] showed the nonlinear disparity in converting Knoop to conventional macro Rockwell C hardness numbers, RCHN, by using a regression analysis program (Fortran II with an IBM 1620 computer). He showed a semilogarithmic relation existed as expressed by the following regression equation —

$$RCHN = b_0 + b_1 \ \log_{10} KHN \qquad\qquad (Eq. \ 6)$$

where, for example, using an indenting load of 200 g–f, $b_0 = -170.92$ and $b_1 = 79.738$. Using the values of b_0 and b_1 obtained from the analysis and the logarithm of the KHN, the conversion to equivalent RCHN can be readily accomplished. The accuracy of the conversion can be shown to be within plus or minus two numbers of the true Rockwell C hardness 97.5 percent of the time.

It has been shown graphically [18] that the KHN jumps up steeply below $P_1 = 100$ g–f when the indentation load is plotted on a log scale versus KHN on a linear scale. The apparent increase in KHN has been observed by others. Fig. 3 shows the same data on a log–log plot using P.N. . . . the increase of KHN is less sudden and seemingly gradual down to $P_1 = 15$g–f. It should be pointed out that the diagonal graph lines are not evenly spaced parallels and the rate of increase in KHN is less pronounced with soft test blocks (RCHN 25) than with hard ones (RCHN 63.9).

CONCLUSIONS

Though the Knoop hardness test has opened new fields in surface and microstructural investigations, the measured values have to be interpreted with caution. Conversions of the different micro and macrohardness numbers are possible within limits but are never straightforward since many parameters of testing and the properties of the materials to be tested can and do influence the test results to a remarkable extent. It would be most desirable to study systematically such interrelations with the goal to employ reliably small indentation loads of 25 g–f or less.

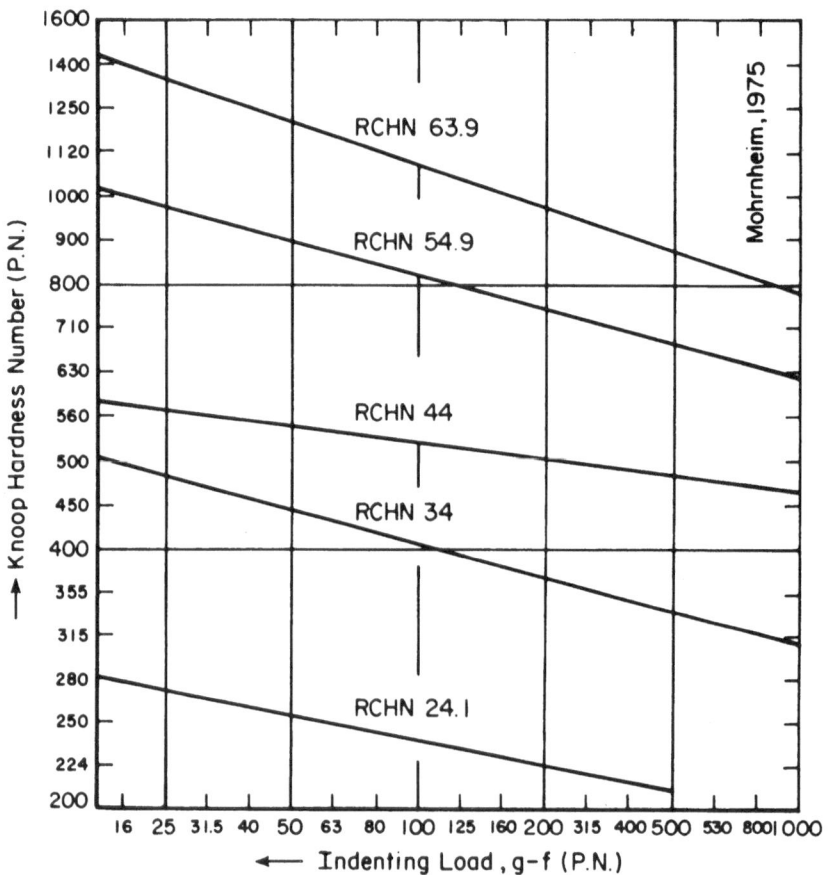

Fig. 3. Increase in KHN with decrease in indenting load for several RCHN levels [18].
Simplified log-log plot featuring preferred numbers (P.N.).

REFERENCES

1. Knoop, F., Peters, C.G., and Emerson, W.G., "Sensitive Pyramidal — Diamond Tool for Indentation Measurements," *Journal of Research,* National Bureau of Standards, v. 23, 45 —46 (1939).
2. Wilson Instrument Division of ACCO, Bridgeport, Connecticut, 06602.
3. *Standard Method of Test for Vickers Hardness of Metallic Material,* ASTM Designation E 92 —67 under the jurisdiction of the ASTM Committee E—1 on Methods of Testing. Replaces E 92—65. (Originally issued 1952.)
4. American Society for Testing and Materials, 1916 Race St., Philadelphia, Pa. 19103.
5. H. Modin and Sten Modin, *Metallurgical Microscopy,* (1968), revised translation, John Wiley and Sons, Inc., New York (1973).
6. *Standard Method of Test for Microhardness of Materials,* ASTM Designation E 38 4—69 originally published under the jurisdiction of ASTM Committee E—4 on Metallography, adopted 1969. Current edition ASTM Designation E—38 4—73 (published in May 1973).
7. *Standard Method for Measurement of Microhardness of Electrodeposited Coatings,* ASTM Designation B—578—73, approved March 1973 by ASTM Committee B—8 on Electro-deposited Coatings and Related Finishes.
8. Research Laboratory, Palo Alto, California through the Northeast Academic Science Information Center, NASIC.
9. Hays, C., "Analysis of Knoop Microhardness," *Metallography,* vol. 6, no. 4, 275—282 (Aug. 1973). Modified version of Kick's law, prove theory valid.
10. Fett, Th., Nothdurft, W., and Racke, H.H., "Measurement of Knoop Hardness for Determining of Orientation of Amorphous Thermoplastics," *Kunststoffe.*
11. "Determination of Yield Locus Curves for Copper and Aluminum by means of Knoop hardness Measurements," *Z Metallkd,* vol. 63, no. 10, 618—822 (Oct. 1972).
12. Egorov V.M., Malysher A.I. , and Fedotov, A.I. "Hardness Determination on the surface layers and films with the Knoop Indenter," *Zavod Lab,* vol. 37, no. 5, 601—604, (1971).
13. Dengel, D. and Rossow, W.E., "Investigation of the Compatibility of Vickers and Knoop Hardness Values," *Haerterei Tech Mitt,* v. 26, no. 1 (April, 1971).
14. Racke, H.H. and Fett, T., "Determination of biaxial internal stresses in the surfaces of plastics by Knoop Hardness Measurements," *Materialpruefung,* vol. 13, no. 2, 37—42, (Feb. 1971).
15. Wonsiewicz, B.C. and Wilkening W.W., "Comparison of conventional and Knoop hardness yield loci for magnesium alloys," *Met Soc of AIME Trans* vol. 245, no. 6, 1313—19 (June 1969).
16. Shiraishi, M., Kimura, H., Jumai, J. and Yoshida, Y., "Temperature variation of coal," *Fuel Soc. Japan—J,* vol. 47, no. 497, 695—701 (Sept. 1968).
17. Batchelder, G.M., "The nonlinear disparity in converting Knoop to Rockwell C. Hardness," *Materials Research & Standards,* 27—30. (Nov. 1969).
18. *Standard Hardness Conversion Tables for Metal,* ASTM Designation E 140 — 67 Part 31, 564, (1969).
19. Mohrnheim, A.F., "Square Ruled Charts and Preferred Numbers for Plotting and Evaluating Wire Drawing Variables," *Wire & Wire Products, 37,* 193—198 (Feb. 1962).
20. Mohrnheim, A.F., "Short Notes — Vickers and Knoop Hardness Numbers of Metallic Materials," *Practical Metallography,* vol. 10, 94—97 (1973).

21. Meyer, K., "Attainment of Comparability of Hardness Measures, a Problem of Materials Testing of Instructional Significance," VDI Ber. 11, 103–122 (1957).
22. Mohrnheim, A.F., "Numerical Relations Between Grain Size, Strength, and Hardness," *Practical Metallography* vol. 5, no. 3, 144–149 (1968).

NON-AMBIENT TEMPERATURE MICROSCOPY

ROBERT Z. MUGGLI and WALTER C. McCRONE *

INTRODUCTION

One of the authors has been quoted as saying that any industrial process can be duplicated under the microscope. To achieve this requires, among other things, wide ranges of controlled temperature, both below and above room temperature in the specimen area. Temperature itself should not be a problem, however, since objects from liquid helium temperatures (-268.6°C) to flame temperatures (about 4500°C) can be studied microscopically. Even high DC arc and spark temperatures are available but, especially with flames, arcs and sparks, temperature control and read-out become a problem.

Materials of construction and protection of microscope optics also become problems above the range of commercially available hot stages (1600°C). Most microscopists construct their own hot and cold stages when their needs fall outside the -50 to $+1600^{\circ}$C range. Thermocouples can be used for temperature sensing up to about 3000°C (W/W—Re) and optical pyrometers beyond that point. Thermistors and thermometers are, of course, often used near room temperature and platinum resistance thermometers can be used up to about 1500°C.

Two unusual high temperature thermometers based on gallium and tin also have been used. Gallium with a liquid range from $29.7 - 1600^{\circ}$C has been used in a quartz capillary thermometer over the range $30 - 1000^{\circ}$C. Tin, with a liquid range from $231.8 - 2260^{\circ}$C, has been used in a graphite thermometer using a capacitance clip around the graphite to detect the molten tin meniscus; it is useful to about 1700°C. Thermocouples and platinum resistance thermometers have pretty much replaced these more heroic thermometers.

ELECTRON MICROSCOPES

Although materials can be studied at non-ambient temperatures with both optical and electron microscopes, the temperature range for the latter is, at least until now, only -268.6 to $+1100^{\circ}$C. There seems no reason, however, why temperatures up to the melting point of tungsten (about 3400°C) could not be studied with the electron microscope.

---

*McCrone Associates, Chicago, Illinois, USA.

Cold Stages

A number of devices have been developed to alleviate specimen contamination in the electron microscope caused by condensation of carbon from organic vapors on the specimen. These are usually very cold metallic condensing surfaces arranged near the specimen so that, hopefully, contamination will condense thereon and not on the specimen. It is a simpler matter than to place the specimen in contact with such a cold surface in order to obtain lower than ambient temperatures. Schott and Leisegang [1] described one of the earliest cold stages for the electron microscope. Although it too was designed to reduce contamination it did lower the specimen temperature to about −120°C. Contact between the specimen cartridge and a liquid nitrogen-cooled copper rod effected the cooling.

A more recent system designed by Anderson and Lucas [2] utilized a liquid nitrogen reservoir in the specimen stage. A number of liquid helium stages also have been built and Valdre [3] succeeded in building such a stage with specimen tilting capability. A completely different approach to the problem of examining very cold specimens with the electron microscope, and in many cases fulfilling much the same purpose with less trouble, is to replicate the chilled specimen before placing it in the microscope. The replica itself is then examined at ambient temperature.

Commercial cold stages are available for most of the transmission electron microscopes. Philips, Hitachi and Zeiss, among others, furnish cold stages generally used down to about −160°C. Fig. 1a shows the cooled sample grid holder for the Philips electron microscope.

Hot Stages

There are two approaches to the examination of heated specimens by electron microscopy. The usual procedure is to heat the specimen in a conventional transmission or scanning electron microscope. A unique solution to the problem, however, pioneered by Rouze and Grube [4], is to build a thermionic emission electron microscope in which the specimen becomes the electron source. Such an instrument can only be used in the thermionic emission range above about 400–500°C, but its upper limit depends on the specimen. The resolution is only about 50 nm, far short of the <0.5 nm of the conventional transmission electron microscope. However, the ability to look at thick specimens and the excellent specimen contrast due to composition and orientation variations in the thermionic emission and the real-time look at phase changes, grain growth, etc. helps compensate for the lack of resolution.

Transmission electron microscopy at high temperatures requires thin metal foils placed on a heating platform [5]. The annealing of aluminum has been studied by Silcox and Whelan [6] using a special specimen cartridge, stage assembly and retractable rod assembly designed into an Elmiskop I. Agar and Lucas [7] have improved on this system using a standard TEM grid and avoiding interference between heating current and electron beam. Heating currents of two amps give 1100°C almost instantaneously.

Small local cartridge heaters are now available for most electron microscopes. These have advantages of more uniform specimen temperatures and approximate temperature

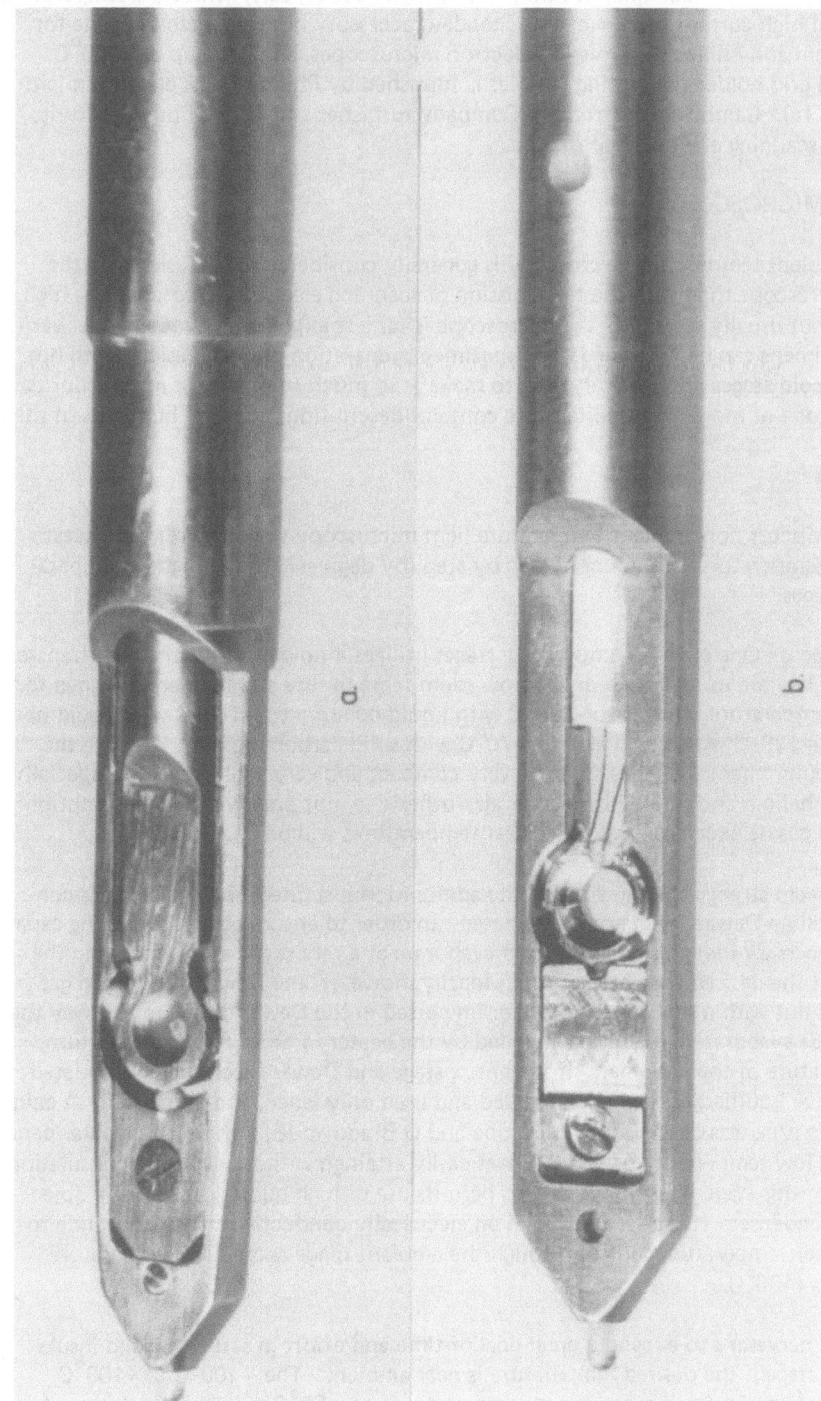

Fig. 1. Special sample grid holders for the Philips transmission electron microscope: a) cooling stage, b) heating stage.

measurement by thermocouple, but they have a high heat capacity with slow temperature changes and high current requirements. Standard accessory hot stages are available for Zeiss, Hitachi and Philips transmission electron microscopes, all usable up to 1000°C. A modified grid holder for heating samples is furnished by Philips for its electron microscope (Fig. 1a). Cambridge Instrument Company furnishes an 1100°C hot stage for its Stereoscan scanning electron microscope.

OPTICAL MICROSCOPY

Non-ambient temperature microscopy is generally considered to be easier with the optical microscope than with the transmission or scanning electron microscopes; a high vacuum is not usually required; the microscope is far less expensive; a much wider variety of specimens can be examined; and specimen preparation is much easier. Both hot stages and cold stages are relatively easy to make — so much so that most microscopists have made one or more and the literature contains descriptions of many hundreds of them.

Cold Stages

We will discuss non-ambient temperature light microscopy starting with the lowest attainable temperatures and proceed step by step (by degrees?) to the highest temperature hot stages.

Cold gas stages. One class of temperature stages utilizes a flowing gas as the heat transfer medium. They are usually used only below room temperature and they are designed for different temperature ranges: to−260°C with liquid helium; to −185°C with liquid nitrogen; to −170°C with liquid air; to −70°C with solid carbon dioxide. All have the common requirement that they must be very compact and very well insulated, especially with liquid helium and liquid nitrogen or air, otherwise, not only will the consumption of liquified gas be excessive, but the lowest temperatures will be unattainable.

It may seem strange that, in spite of the admonition just cited, the liquified gas container, usually a Dewar, need not be insulated. In order to ensure adequate cooling capacity it is necessary that this cooling liquid evaporate at a rate sufficient to maintain the specimen at the desired low temperature. Ideally, however, the Dewar of liquified gas is insulated but with a small electric heater immersed in the Dewar itself. In this way the liquid can be evaporated at a rate controlled by the heater in order to control, in turn, the temperature of the specimen. If the entire stage and Dewar is completely insulated the expensive liquified gas can be conserved and used only when, and as, needed. A cold stage of this type was described by McCrone and O'Bradovic [8]. In that particular paper the desired low temperature of −100°C was easily attained without very much insulation. A feature of this stage is its thinness; this permits use of high numerical aperture objectives and condenser. It is also fitted with an electrically conductive microscope slide to permit better temperature control through the ambient range and a high temperature range up to +100°C.

It is not necessary to expend a great deal of time and effort in setting up and insulating a cold stage if the desired temperature is near ambient. The −100°C to +100°C stage was designed primarily to cover the range down to −50°C with the capability of permitting phase contrast illumination. At that time, this meant a very thin stage so that the phase condenser annulus could be properly imaged in the objective back focal plane.

To attain the lowest possible temperatures every possible precaution must be taken to avoid heat loss — short gas lines and abundant efficient insulation (Fig. 2). The more attention one pays to these requirements the closer the lowest temperature will be to the boiling point of the liquified gas in use.

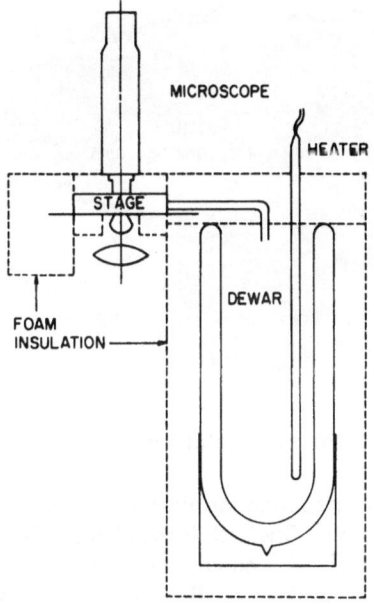

Fig. 2. Arrangement for efficient cooling of a cold stage using liquid air or nitrogen.

There are, to our knowledge, no commercially available stages of the type described above. There is, however, a simple Leitz cold stage (Heating and Cooling Microscope Stage 80; Fig. 3) that could be insulated for use at temperatures lower than the advertised −20°C. It also can be used with any standard microscope optics.

Joule-Thomson cold stages. Another class of cold stage utilizes liquid helium, hydrogen, argon, nitrogen or air, but the respective gas is liquified in the stage itself by a Joule-Thomson effect. The temperature required determines the specific gas selected. Temperatures at or near the boiling point of the liquid can be attained in a well-designed Joule-Thomson cold stage. The significance of the inversion temperature is that the gas cannot be liquified by a Joule-Thomson expension if its starting temperature exceeds that figure. Argon, air and nitrogen therefore can be liquified starting with the gas at room temperature. Hydrogen, however, must be precooled to −69°C before expansion and helium must first be cooled to −253°C. It is, however, relatively simple to precool gaseous hydrogen with liquid air or liquid nitrogen before Joule-Thomson liquifaction of the hydrogen. Helium, on the other hand, would require pre-cooling by hydrogen before it could be liquified.

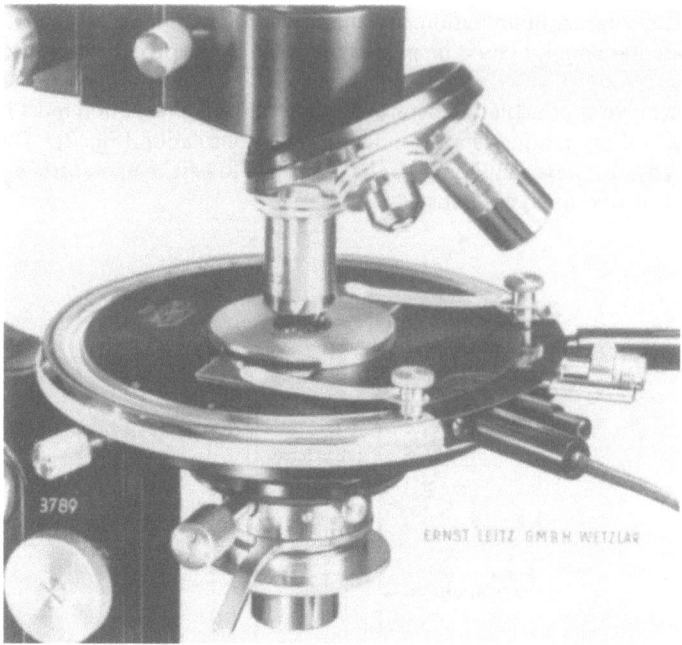

Fig. 3. Leitz 80 stage useful from −20°C to +80°C.

Although cryogenic cold stages sound formidable to most microscopists, the devices available are relatively easy to use. One such is the Hymatic Minicooler [9], which is marketed in the U.S. by Bendix*. A second essentially identical device is marketed by EMI in England under the name Emicooler. Both devices conduct the gas through finned metal capillaries in such a way that the gas exiting through the final 20μm orifice flows back over the fins to precool the incoming gas.

With a flow rate of about 0.4 cfm of gas at an initial pressure of 4000–6000 psi, liquified gas can be observed at the orifice in a few seconds. These devices have one disadvantage due to the tiny 20 μm orifice; they clog easily with tiny particles or with solid CO_2 or ice. It is necessary to very completely remove these materials from the entering gas stream to ensure continuous trouble-free operation. We have found that a molecular sieve serves both as a gas absorber and as a particle filter. Since we also usually use oil-pumped nitrogen in large cylinders, there is little cleanup required and the molecular sieve has a long life. Any such assembly of tank, pressure regulator, high pressure line and filter should be carefully and forcefully blown out with the feed gas before connecting the cooler.

The Minicooler as furnished is best adapted for microscopical study of opaque objects or, at least, by reflected light. Bendix will, however, provide other configurations so that transmitted light can be used. The Emicooler can be used either for transmitted or reflected light although the polymer used for the stage is not isotropic nor can it usually

* available in the U.S. from Bendix Instruments and Life Support Division, Davenport, Iowa 52808

be turned to an extinction position common to both top and bottom plates. A small thermocouple bead is easily imbedded in either stage just below the specimen. The latter should be mounted on the stage plate itself or on a tiny fragment of coverslip.

To work at liquid hydrogen temperatures ($20°K$) requires precooling of the hydrogen gas using a Dewar of liquid air. The hydrogen passes through a copper coil immersed in the liquid air immediately before the cooling stage. Again, adequate insulation and short lines are essential.

Peltier devices. The Peltier effect has been known for many years and is the basis of the common thermocouple. Two dissimilar metals in contact will develop an electrical potential between them if the junction is heated or cooled. The converse is also true, but with any ordinary metal the effect is too small to be useful in a microscope cold stage. However, the relatively recent development of semiconductors has made this application of Peltier cooling practical. To accomplish this, a junction is formed from two semiconductor crystals, one a p-type and the other an n-type. When a current passes across this junction either heating or cooling occurs, depending on the direction of the current. The degree of heating or cooling is a function of the current being carried and the resistance of the semiconductor.

Usually, single crystals of bismuth telluride or antimony telluride are used as the semiconductor and the current requirements range up to 25—30 amps at 1—2 volts. Obviously, at these amperages resistance heating is a limiting factor. Most thermoelectric junctions in use will produce a maximum temperature difference, under optimum conditions, of at least $65°C$. Since one end of each crystal becomes hot and the other cold, one must prevent conduction of heat from the hot end to the cold end by circulating cold water in contact with the hot junction or simply with copper cooling fins. A more elegant way to cool the hot junction is, of course, to use a second thermoelectric cooling unit. One could consider cascading several more units to thus gain additional cooling. It turns out, however, that very little additional cooling is effected beyond 2—3 units and, besides, each additional cooling unit must be larger than its predecessor. This, plus the power supply problem, usually limits the design to a 2—unit system.

The system we use [10] has one unit of 2 p— and n—pairs cooled by a second unit containing 8 p— and n—pairs. Units containing 8 pairs can be purchased at relatively low cost already assembled. Our 2—stage unit has a cooling water coil to remove heat from the lower hot junction and the entire assembly is insulated with plastic foam in a small box fashioned from polystyrene. The microscope objective fits through an O-ring lining a circular opening cut in the center of the polystyrene top, thus sealing the stage from easy access to moisture in the air which would condense on, and obscure, the preparation.

A 2 mm opening is drilled vertically through the centers of the two thermoelectric units taking care to avoid the semiconductor crystals; this forms the light well so that transmitted light may be used.

This cold stage has been very satisfactory, particularly because one need only add a liter of water to the cooling bath and turn on two switches to start; within three minutes the stage temperature is down to $-40°C$. Cooling to this low level requires no longer than the usual time of preparing the sample for observation and focusing the microscope.

Any desired temperature from $50°C$ to $+100°C$ can be quickly set and held by manipulation of the variable voltage and the direction of the current flow. Either controlled heating and cooling are easily managed and equilibrium melting points in which the crystal and melt are maintained in dynamic equilibrium are easily carried out.

Carbon dioxide stages.   The temperature range for commercially available stages begins at about $-50°C$ although some can be used at lower temperatures by using lower boiling liquids as coolants. Both Reichert and Leitz produce stages designed for carbon dioxide. The Leitz Heating and Cooling Microscope Stage 80 is shown in Fig. 3. It also has facilities for heating so that positive temperature control is possible near room temperature. Reichert makes the Kofler cold stage which is similar to the Kofler hot stage but with a circulating gas chamber below the specimen stage. A heating unit, limited to $+80°C$, is also incorporated to give positive response near ambient temperature.

Of course, any circulating gases at any temperature can be used with these stages. Carbon dioxide is convenient and gives specimen temperatures down to about $-60°C$.

Circulating fluid stage.   A simple glass stage for use near, but either above or below, room temperature is shown in Fig. 4 [11]. It can be used with any laboratory liquid pump and any of the common provisions for heating or cooling liquids. Actually, quite precise temperature measurements are possible with this stage since temperature equilibrium is so rapid. A glass blower can usually fashion the stage itself in about 10–15 minutes. The top surface should be as flat as possible. The preparation, on a half-slide, is "sealed" to the top plate of the glass stage with several drops of glycerin or ethylene glycol to improve both light and heat conduction to the preparation. The stage may be insulated if temperatures much below zero or above about $100°C$ are used, but otherwise this is unnecessary.

Mettler FP–5 as a cold stage.   The Mettler FP–5 , discussed below under hot stages, can be used as a cold stage down to $-20°C$ by circulating cooled gas into the fan system. The digital temperature readout functions over the range -20 to $300°C$, hence positive temperature control over the entire range can be obtained by having a base temperature of less than $-20°C$ in the cooling gas. If short gas lines and good insulation are used dry nitrogen gas passing through a copper coil in a dry ice cooling bath can be used effectively. Mettler will, in fact, modify their stage so that the range becomes -50 to $270°C$ (or 30 to $350°C$) on request.

Quick-chill aerosols.   Several devices to chill cocktail glasses are now being marketed as bar accessories. The inexpensive refills for these are small pressurized, aerosol cans containing liquified gases such as dichlorodifluoromethane. The discharge from these cans lowers the temperature of a microscope slide to $-50°C$ in a few seconds.

Hot Stages

As with cold stages, the commercial hot stages cover the range near room temperature.

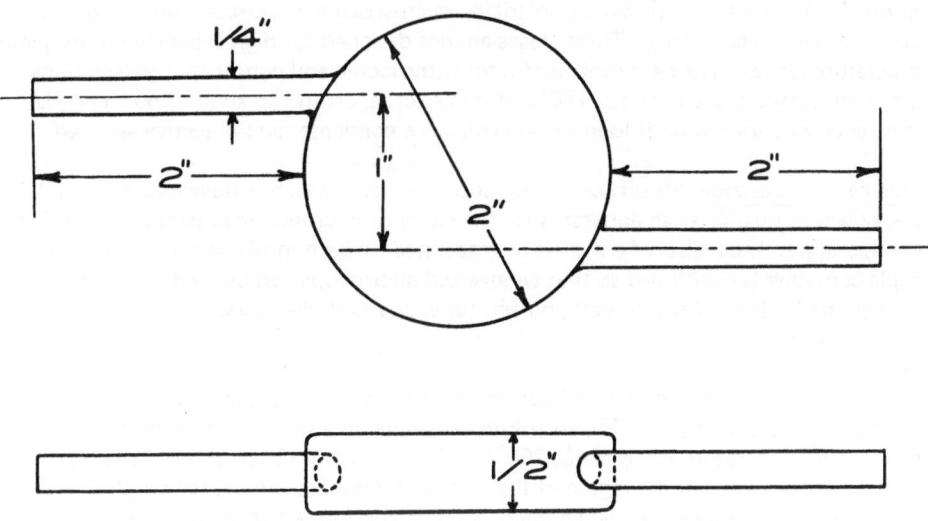

Fig. 4. A simple circulating fluid stage.

Above about 1500°C microscopists must make their own stages and many even prefer to make their own stages in the lower temperature range. Possibly the largest single group of papers on microscopical topics is the one covering hot stages.

Probably the first hot stage was a strip of copper extending over the edge of the micro-scope stage so that it could be heated with a candle or micro gas flame. A specimen on the other end of the strip could then be observed microscopically. Later, Otto Lehmann introduced a small gas flame immediately under the center of the stage to heat directly a metal sample holder. Since then, of course, hot stages like nearly everything else have be-come more sophisticated, versatile and expensive.

Hot-wire stage. Still relatively simple are the hot-wire stages. An electrically heated wire held just above a microscope coverslip can be used to impose a temperature gradient on any specimen. Jones [12] has used such a device to study composition diagrams and Hartshorne [13] more recently has applied an improved device to the study of phase transitions. He has been able to make quite accurate (±1−2°C) measurements of phase boundaries by calibrating the system with a compound having 2−3 known phase transitions. Two preparations, one known and one unknown, are arranged side by side in the field of view of a low power objective (30−50 mm focal length) and the positions of the transitions for each on an eyepiece micrometer scale are recorded. Plotting of the known transition data against eyepiece scale divisions gives a calibration curve for the unknown. Hartshorne finds that the plot of $\log(t-t_r)$ vs S is a straight line. S is the distance of the transition from the wire, t is the temperature at the transition and t_r is room temperature.

Thin-metal stage. Hartshorne [14] also has designed two thin-metal stages: a rectangular one for the Dick Model (Swift) polarizing microscope and a circular one for circular stage polarizing microscopes. These stages are not designed for high accuracy nor extreme temperature ranges. They are most useful for orthoscopic and conoscopic observations from room temperature to about 150°C at numerical apertures to about 0.50. This NA can be exceeded somewhat if long working distance condenser and objective are used.

Oil bath microscope. It is difficult to describe this setup as a hot stage microscope. It does, however, qualify as an apparatus for the study of microscopic objects as a function of temperature. It consists of a heated thermostated oil bath in which a long flat-bottomed sample container is positioned so that an inverted microscope can be used to observe (through the bottom of the oil bath, oil and tube) any particles suspended in a liquid inside the tube.

Its special purpose is the study of polymorphic transformations and measurement of transition temperatures [15]. Excess solute suspended in a saturated solvent can be agitated at any temperature up to about 200°C with continuous observation of the crystalline phase present. In the region of the transition temperature the temperature can be held constant to make sure which phase is stable and the temperature at which two phases are in equilibrium with the saturated solution, i.e., T_{tr}.

Electrically coated microscope slide. If a microscope slide is coated with a conductor of electric current but with a reasonable electrical resistance a current can be passed through the coating and thereby heat the slide. Such electrically coated (EC) slides can be prepared by evaporating onto the slide a very thin metal coating, e.g., platinum or by hydrolyzing tin chloride with steam on the glass surface. The latter gives the most inert and stable coatir It has been used to make heated windshields and other heating equipment or elements. We use these EC slides as heating elements in the −100 to +100°C cold stage [8]. The EC slides are also useful as low temperature heaters by themselves on a microscope slide. A thin strip of conducting paint on either end of the slide serves as an electrical contact for two wires going to a variable voltage control.

Kofler hot stage. Thousands of Kofler hot stages are in use throughout the world today. Developed by Ludwig Kofler in 1931 (Fig. 5), this well-engineered stage has been responsible for making hot stage methods a science rather than a hobby. Only when the Mettler hot stage appeared nearly 40 years later with digital temperature read-out, programmed heating rates and no vertical heating gradient did the Kofler hot stage suddenly become outclassed. Although the Kofler stage costs nearly an order or magnitude less than the Mettler stage, the latter has become the instrument of choice for those who can afford it. A careful microscopist can still, however, do with the Kofler stage everything a good microscopist can do with the Mettler. Operations, routine with the Mettler, require more skill and care with the Kofler and often 2−10X the work time.

The Kofler stage can be purchased from C. Reichert, William A. Hacker or Arthur H. Thomas in essentially identical designs covering the range from room temperature to 350−360°C. Arthur H. Thomas furnishes the stage either with thermistor read-out or two thermometers (30−230°C and 60−350°C). Although heat baffles are furnished, the Kofler stage like all other stages (including the Mettler) must be carefully calibrated [16].

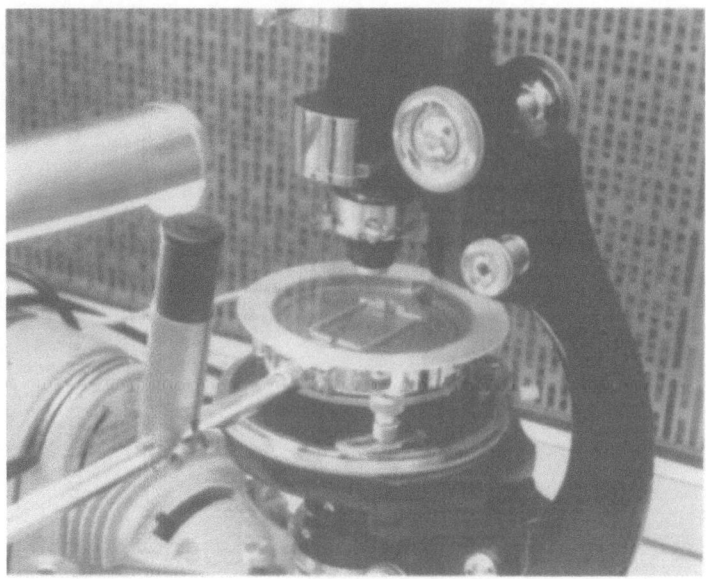

Fig. 5. The Kofler hot stage as furnished by Arthur H. Thomas Company.

Mettler FP–5 hot stage. The Mettler hot stage (Fig. 6) is a superb instrument. Many of the difficulties encountered with the Kofler and other hot stages are due to the vertical temperature gradient. The Mettler stage gives a uniform temperature through the preparation volume by sandwiching it between two identical heating elements. A fan blows cool air around the outside of the heating elements and inside the case to keep the case cool and to cool the stage more rapidly. A platinum resistance thermometer is embedded in the lower heating element to ensure accurate temperature sensing. A variety of linear heating rates from 0.2–10°C/min are available over the range from −20°C to +300°C. A reversing switch permits the same range of linear cooling rates. The temperature read-out is digital and there is a memory bank for remote recording of three temperatures during an experimental run.

It is unfortunate that microscopists tend to believe digital read-outs implicitly because the numbers obtained with the Mettler, or any other hot stage, are seldom if ever correct. There is, in fact, no way that any hot stage can be accurate under all experimental conditions. Every hot stage must be calibrated for the particular experimental conditions if accurate results are to be obtained [17].

Leitz −20 to +350°C stage. Leitz furnishes a stage equivalent to the Kofler stage and with several special features. It can be placed on almost any polarizing microscope but it is also furnished with a simple stand especially suited to its use. One feature is a "drawing camera" eyepiece designed to superimpose any portion of the thermometer in the field of view for easy reading of temperatures. It also has a cooling chamber for carbon dioxide or other cold gases or liquids. The three thermometers cover the ranges: −20 to +110°C, +100 to +230°C and +220 to 350°C.

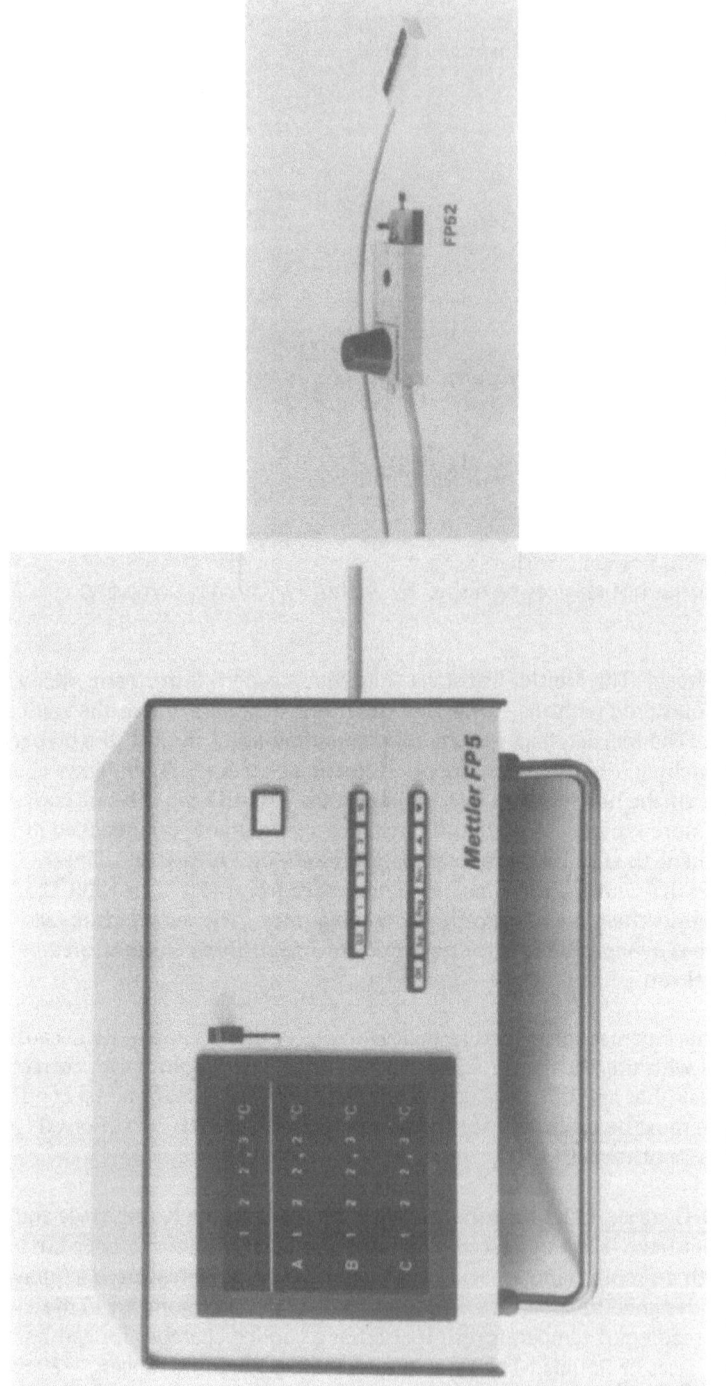

Fig. 6. The Mettler FP—5 temperature programmed and temperature read-out module (left); The FP—52 stage is shown on the right at about two-thirds the scale of the FP—5.

Aminco Accumelt. This is a unique instrument that heats a small (<3 mg) capillary sample at a rate of about 250°C/min. The temperature is recorded by a strip-chart recorder as the response to a thermocouple imbedded in the sample. The total range of 0–300°C is covered in three stages: 0–100°C, 100–200°C and 200–300°C. Melting points and other solid state phase transitions can be read to ±0.5°C in most cases. Although not a microscopical procedure it solves very nicely one of a microscopist's problems: obtaining accurate melting points on decomposible compounds before they have time to decompose.

Stanton-Redcroft 1000°C stage (HMS–5). An excellent stage with provision for linear temperature programming (1°–100°C/min heating or cooling) is furnished by Stanton–Redcroft* (Fig. 7). The stage itself can be replaced with DTA or TGA systems in order to take full advantage of the temperature control unit. See Charsley and Kamp for full details [18]. A unique feature of the stage is the use of reflected light measurement as a function of temperature to show phase changes.

Leitz 1350°C stage. Maria Kuhnert–Brandstätter [19] has done an excellent job of evaluating the Leitz 1350°C hot stage (Fig. 8). She finds it to be reasonably accurate (±5–10°C) if frequently calibrated and if used with an inert atmosphere, preferably argon. She reports that nitrogen as an "inert" gas reacts with some substances when heated, for example, barium chloride gives barium nitride. Other problems to be faced such as composition of slides and covers are also discussed. A major disadvantage of the stage for transmitted light microscopists is the inability to move the preparation once in the stage. A two-mm circle, the area of the light well, is the total preparation area.

Reichert 1600°C stage (Vacutherm). A stage similar to the Leitz 1350°C stage is produced by C. Reichert [20]. It uses molybdenum, tantalum or tungsten heating elements and is restricted to the examination of opaque specimens. This stage has, more recently, been modified for vacuum operation instead of an inert gas. It is now called the Vacutherm.

Union 1800°C stage. The Union Optical Company of Tokyo, Japan, produces a 1800°C reflected light vacuum stage similar to the Vacutherm. They note, however, that temperatures to 2300°C can be attained using optical pyrometry for temperature sensing. A vacuum of 10^{-6} Torr is claimed as well as a 15 second heating rate to 1800°C.

Metal ribbon stages. The idea of electrically heating a thin metal strip doubling as a specimen support and heating element is capable of infinite variation. The metal itself may be varied as well as the power supply and housing. The choice of metal depends not only on its melting point but on its oxidation stability. Most metals except the noble elements like platinum, iridium and gold oxidize in air on heating. Rhenium, for example, can be heated in air to 2000°C or somewhat higher before beginning to oxidize; tungsten oxidizes even more readily as does graphite. If it is necessary to exceed about 2500°C with, say molybdenum, it is necessary to operate either with an inert, preferably argon, atmosphere or with a good vacuum.

---

*Copper Mill Lane, London SW17 0BN, England.

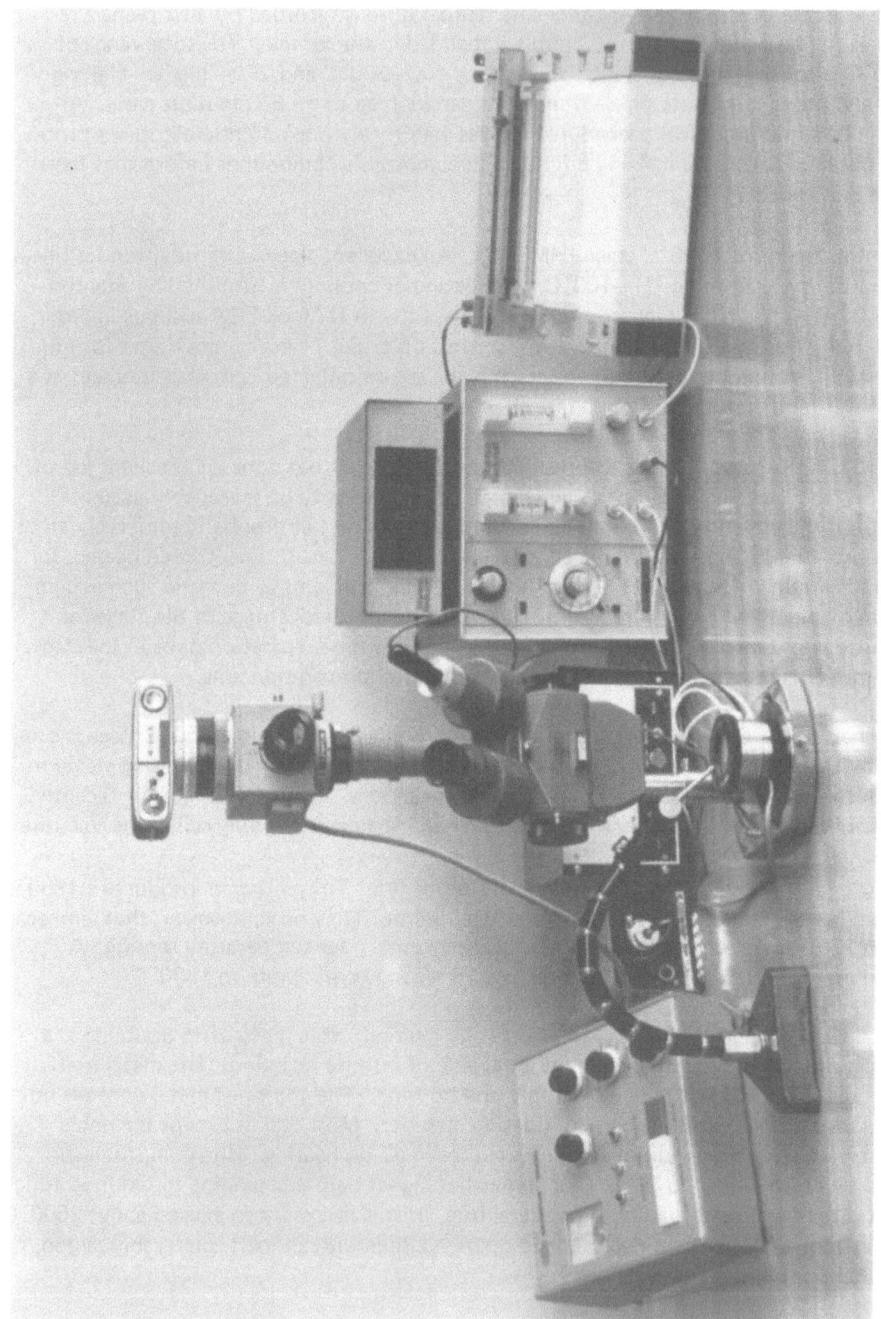

Fig. 7.   The complete assembly for the Stanton—Redcroft 1000°C stage.

Fig. 8. The Leitz 1350°C hot stage for transmitted and reflected light.

Jack Dodd has designed an interesting, yet simple, 2000°C stage. The heating element is a tiny wire loop which acts also as a resistance thermometer. Platinum or iridium are ideal since they do not require an inert atmosphere. The loop itself is so small that the amount of heat is minimal. and almost any containing material can be used. He used a machined lavite block which can be easily drilled to permit vacuum or inert gas operation. The sample is held in the wire loop mechanically or, once melted, by surface tension. The stage is simple to operate and an equilibrium melting point on a single crystal of quartz was easily measured at 1721±3°C. An advantage of Dodd's stage is that the null detector used to measure the resistance of the loop becomes more accurate at high temperatures, i.e., the null becomes sharper. The accuracy therefore, in effect, remains constant at about ±3°C over the entire temperature range.

Alison Instruments* have marketed a very fine 1800°C hot stage related in principle to Jack Dodd's platinum loop device. Both designs use the heating unit as the temperature sensing element: Dodd by using the platinum heating loop as a platinum resistance thermometer; Alison by making the temperature sensing thermocouple a resistance heating loop. Both cycle the two functions many times each second. The Alison instrument is a beautifully engineered hot stage system with an integral microscope and digital temperature read-out. The sample and thermocouple mass are so small that only 3—4 watts are needed for heating to 1800°C and cooling time from 1800° 0 to 700°C is only about 0.5 sec. This instrument is the Cadillac (Rolls-Royce?) of the high temperature hot stages (Fig. 9).

High temperature vacuum stage. Mike Bayard of our laboratory was once asked to photograph tungsten metal surfaces at high magnifications (400—500X) under high vacuum at temperatures up to 2500°C. He decided the easiest way to do this was to heat a tung-

---

* Division of Alison Engineering Ltd., Great Yarmouth, England.

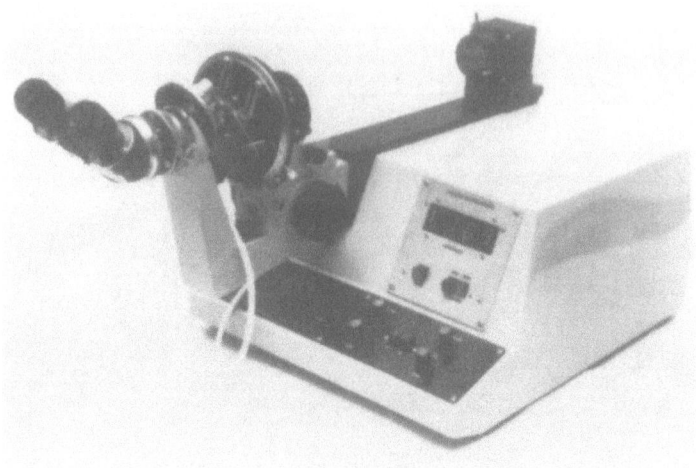

Fig. 9. The Alison 1800°C stage.

sten strip directly and read the temperature by optical pyrometry. A simple microscope
was mounted on the top of a large vacuum stage with a quartz plate as coverslip. A long
working—distance 4—mm objective was used. Many problems arose and were solved before suitable pictures were obtained. One such was clouding of the quartz window by
sublimed tungsten. A magnetically operated "window-shade" solved this problem. One
problem which one might think would not arise is illumination of the self-luminous
specimen. Yet, object detail is not well shown by self-luminosity and Bayard was able
to achieve good detail by taking the picture with a very high intensity Xenon arc, thereby
overpowering the self-luminosity. The vacuum problem was solved with an oil-diffusion
pump; a vacuum of 5×10^{-7} was easily maintained for 2 hours at 2400°C.

The problem of the window composition was solved by choosing silica because of its
low temperature coefficient of expension and high melting point (1728 °C). In other
applications of this stage, we were faced with the need for other high melting inert windows. We found that polished sapphire plates cut from artificial boules were satisfactory
up to about 1800°C; they are, however, birefringent. Later an almost universally suitable material was found, crystalline cleavage plates of periclase (magnesium oxide). These
cleavage plates are isotropic, clear, colorless and melt at 2800°C. We obtained a good
supply as a special favor from Gordon Finlay, then at the Norton Company in Niagara
Falls, Ontario.

Transfer lens system. When the total amount of heat is small, even at ultrahigh temperatures, there is no great risk to the microscope optics or stage materials. Some hot
stages do generate sufficient heat to damage microscope objectives. A solution to this
problem is to use a transfer lens to "look at" the hot object and form an image on which
the microscope objective can be focused. The resolution of such a transfer lens is limited
by its aperture, hence it should be a short focus, low f—stop system. We have used
camera or projector lenses of almost any set of characteristics. A wide angle, say 25—mm

focal length f:1.2 lens operating 50 mm from the hot specimen will deliver a cool image 50 mm on the other wise of the lens.

An interesting application of this kind of system is the work of Henry Bauman at the Carborundum Company. Using a transfer lens made especially for his work by the American Optical Company, he was able to study crystallization of titanium dioxide from a glass, furnace refractory lining dissolving in glass melts, melting of aluminum and mullite, sublimation of alumina, boiling of silica and synthesis of silicon carbide from silica sand and carbon. He used a resistance wound micro furnace for temperatures up to 1500°C and a conduction-heated graphite rod with an opening therein as a cell for higher temperatures up to 2700°C.

Although we know of no microscopy at temperatures higher than 2700°C, there seems no reason to doubt it could be done whenever the need arises.

REFERENCES

1. Schott, O., and S. Leisegang, *Proceedings of the First European Regional Conference on Electron Microscopy,* Stockholm, 20, (1956).
2. Anderson, K., and J.H. Lucas, Paper read at Electron Microscope Group Meeting of Institute of Physics (November 1964).
3. Valdre, U., *Proceedings of the Third Regional Conference on Electron Microscopy,* Prague, 61, (1964).
4. Rouze, S.R., and W.L. Grube, *Proceedings of International Conference on Microscopy 1960,* Chicago, edited by Walter C. McCrone.
5. Whelan, M.J., *Proceedings of Fourth International Conference on Electron Microscopy,* Berlin, 96 (1958).
6. Silcox, J., and M.J. Whelan, *Structure and Properties of Thin Films,* John Wiley and Sons, London, (1959).
7. Agar, A.W., and J.H. Lucas, *Proceedings of Fifth International Conference on Electron Microscopy,* Philadelphia, paper E–2 (1962).
8. McCrone, W.C. and S.M. O'Bradovic, *Anal. Chem. 28,* 1038 (1956).
9. Nicholds, K.E. *et al.,* Hymatic Cryogenic Symposium (November 1973).
10. Markussen, J., and W.C. McCrone, *Microscope 14,* 395–402 (1965).
11. McCrone, W.C., *et al., Anal. Chem. 18,* 578 (1946).
12. Jones, F.T., *Microscope 16,* 37 (1968).
13. Hartshorne, N.H., *Microscope 23,* 177–190 (1975).
14. Hartshorne, N.H., and A. Stuart, *Crystals and the Polarizing Microscope,* Arnold, London, (1970).
15. Teetsov, A., and W.C. McCrone, *Microscope 15,* 13–29 (1965).
16. McCrone, W.C., *Fusion Methods in Chemical Microscopy,* John Wiley and Sons, New York, NY, (1957).
17. Julian, Y., and W.C. McCrone, *Microscope 19,* 225–234 (1971).
18. Charsley, E.L., and A.C.F. Kamp, *Thermal Analysis,* Vol. 1, Proceedings Third ICTA DAVOS (1971).
19. Kuhnert–Brandstätter, M., *Microscope 16,* 257–265 (1968).
20. Gabler, F., and W. Wurz, *Metall. 9,* 819–823 (1959).

REMOTE METALLOGRAPHY

J. H. EVANS *

INTRODUCTION

Nuclear power stations now contribute a significant part of the electrical energy requirements of many countries and it is likely that this contribution will rise in the future. The successful development of a nuclear power program depends largely on the testing of candidate fuel and structural materials in a test reactor and subsequently full scale components from power reactors. After irradiation in a nuclear reactor materials become intensely radioactive and, in order to protect workers, have to be handled remotely in special enclosures. Although hazards involved in handling radio-active materials are formidable, safe remote handling techniques have been developed and operated in many laboratories around the world for over 30 years. Post-irradiation examination (PIE) of irradiated materials involves most of the techniques normally used in conventional laboratories such as radiography, non-destructive testing, chemical analysis and metallography. This paper describes recent developments in facilities and equipment used for metallographic examination of radioactive materials and gives details of the preparation techniques used. The adaption of standard electron-optical instruments for radioactive material examination is also briefly described.

RADIOLOGICAL PROTECTION

After irradiation in a nuclear reactor, materials become radioactive and because of the ionising radiations they emit, cannot be handled directly; all operations on such materials are carried out within protective enclosures fitted with suitable handling devices. Radiations against which protection must be provided consist of a and β particles, γ rays and neutrons. Apart from the external hazard of direct radiation, there is the internal hazard from inhalation or ingestion of these materials and protection from both these hazards is necessary. The method of protection depends on the types of radiation present.

(a) Alpha particles (a) are high energy helium nuclei and are stopped by a few centimeters of air or a thin sheet of paper. Alpha emitters are extremely dangerous when inside the body and hence it is essential to contain alpha active materials to prevent ingestion/inhalation, but heavy shielding is not required. Examples of alpha emitters are artificially produced plutonium 239 and uranium 233.

*Metallurgy Division, AERE, Harwell, Oxon, ENGLAND

(b) Beta particles (β) are high energy electrons emitted from the nucleus in certain types of radioactive disintegration and are stopped by thick cardboard or a thin sheet of aluminum. Strontium 90 is a beta emitter.

(c) Gamma rays (γ) are electromagnetic radiations emitted by the nuclei of radioactive substances during decay and are only stopped by thick shields of dense materials such as concrete, lead or cast iron. Examples of gamma emitters are fission products and irradiated reactor structural materials.

(d) Neutrons (η) are penetrating particles and can only be stopped by composite shielding made up of moderating material (such as plastics) a neutron absorber (such as cadmium) and dense material (such as lead or concrete). Californium is a spontaneous neutron emitter.

The type of enclosure necessary to protect workers will depend on the types of radiation present. In the case of alpha active materials only the ingestion hazard has to be considered and handling of such materials is carried out in sealed glove boxes. In cells for handling structural materials the alpha hazard is absent and an alpha box is not necessary. Irradiated materials in which all types of radiation are present i.e. nuclear fuel elements, have to be handled in sealed boxes (to give protection against the internal hazard) surrounded by dense shielding (to protect against the external hazard).

These measures must be adequate to ensure that the radiation level at the working face of a shielded cell does not exceed that recommended by the International Commission on Radiological Protection.

CONTAINMENT DESIGN

Glove Boxes

These are typically constructed of a metal frame with plexiglass panels fitted with gloves. A negative atmosphere is maintained inside the box to ensure all leaks are inward. Several interconnected boxes are used for metallographic preparation and optical microscope examination; each preparation stage is carried out in a separate box. As handling via gloves is possible within the boxes, conventional metallographic equipment can be utilized although specially designed equipment is sometimes necessary.

Optical microscopes for use in a glove box are available from a number of manufacturers, basically they are similar in design and are constructed so that the body of the microscope is inside the box with the lamp and viewing/ camera system outside. Discs of optical glass fitted at suitable points in the microscope prevent leakage of contamination from the box.

Shielded Cells

Cells for handling $\beta\gamma$ materials can be constructed of concrete or lead (cast iron and steel are also used). When dealing with $\alpha\beta\gamma$ active materials, in addition to dense shielding, an inner containment box is necessary to protect against the internal hazard. As in the case of glove box operations the cell atmosphere, which may be argon or nitrogen when handling reactive materials, is maintained at a negative pressure.

In concrete shielded cells the walls, from 900 to 1600 mms thick are made of concrete in a range of densities from 2.5 – 5.5 gm/cc. Handling is carried out by means of master/slave manipulators and viewing of the cell interior is through windows made of glass or zinc bromide solution contained in a tank fitted with glass end plates.

Lead cells are constructed of lead bricks up to 250 mm thick. Handling is by means of fingers attached to the end of tong shafts which are supported in spheres of tungsten or uranium set in the cell walls. This type of cell is generally used to shield optical microscopes; advantages of lead shielding are that optical transfer systems can be kept much shorter and it permits a form of construction in which one wall can be constructed in the form of a door which after removal of the sample can be opened for cleaning and maintenance of the microscope via gloves in the alpha box panel.

The table below shows the approximate thickness of shielding required to attenuate γ radiation to the recommended level at the operating face.

TABLE 1

Approximate Shielding Thickness Required to Attenuate
γ Radiation to the Recommended Level

Activity in Curies*	Wall Thickness (mm)		
	Lead	Steel	Concrete
10	150	250	800
10^3	250	400	1200
10^5	360	530	1600

*Curie is the quantity of a radio-isotope that decays at the rate of 3.7×10^{10} disintegrations per second.

At Harwell both concrete and lead cells are used for remote metallographic examination procedures. A concrete shielded cell with lead shielded optical microscope is shown diagrammatically in Fig. 1; a typical lead cell is shown in Fig. 2. The layout of a Harwell lead shielded metallographic preparation and examination cell used for $\alpha\beta\gamma$ materials is shown in Fig. 3.

The remainder of this paper is concerned with the metallographic examination of materials in shielded cells.

SAMPLE PREPARATION

Procedures for metallographic preparation of irradiated materials are generally similar to those employed in laboratories engaged on examination of unirradiated materials.

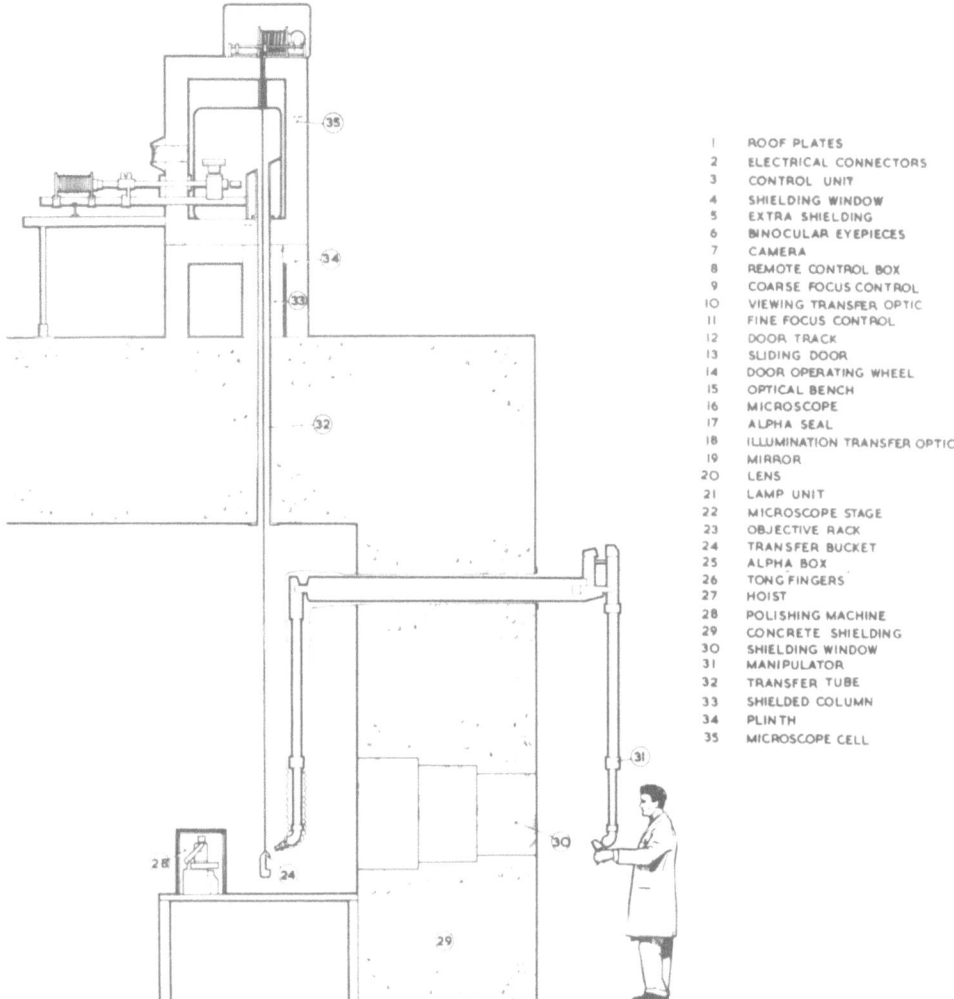

1	ROOF PLATES
2	ELECTRICAL CONNECTORS
3	CONTROL UNIT
4	SHIELDING WINDOW
5	EXTRA SHIELDING
6	BINOCULAR EYEPIECES
7	CAMERA
8	REMOTE CONTROL BOX
9	COARSE FOCUS CONTROL
10	VIEWING TRANSFER OPTIC
11	FINE FOCUS CONTROL
12	DOOR TRACK
13	SLIDING DOOR
14	DOOR OPERATING WHEEL
15	OPTICAL BENCH
16	MICROSCOPE
17	ALPHA SEAL
18	ILLUMINATION TRANSFER OPTIC
19	MIRROR
20	LENS
21	LAMP UNIT
22	MICROSCOPE STAGE
23	OBJECTIVE RACK
24	TRANSFER BUCKET
25	ALPHA BOX
26	TONG FINGERS
27	HOIST
28	POLISHING MACHINE
29	CONCRETE SHIELDING
30	SHIELDING WINDOW
31	MANIPULATOR
32	TRANSFER TUBE
33	SHIELDED COLUMN
34	PLINTH
35	MICROSCOPE CELL

Fig. 1. Lead Shielded Optical Microscope in a Hot Cell

In shielded cell operations however, the metallographer is separated from his sample by up to 3 metres and operations such as cutting off, grinding/polishing and etching have to be performed remotely. Since the time taken to prepare a sample remotely is considerably longer than in a conventional laboratory, the precise procedures used merit detailed attention to minimize this penalty.

The following brief comments on sample preparation are meant to show how standard procedures are adapted for use in shielded cells. Further details can be obtained from previously published papers (1,2).

Sample cut off is carried out with abrasive wheel machines suitably modified to allow remote loading/unloading and wheel changing. A shroud, connected to a high

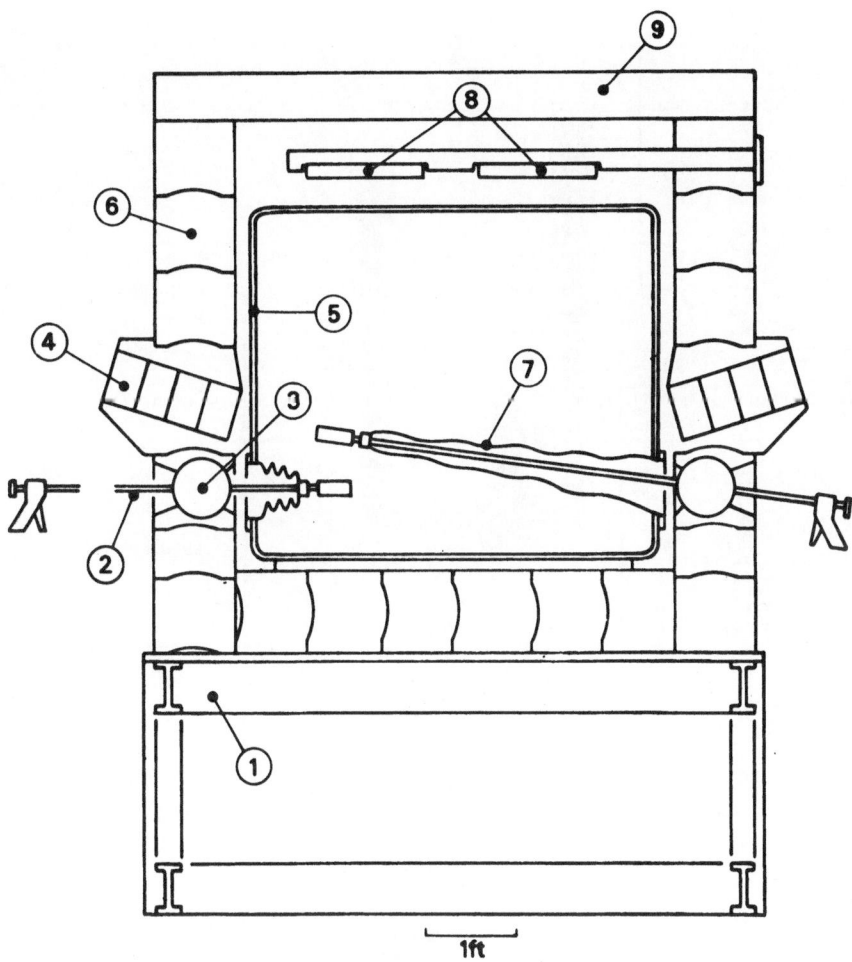

Fig. 2. Lead shielded cell; 1. Plinth; 2. Tong shaft; 3. Sphere unit; 4. Window;
5. "Alpha" box (fitted with Perspex panels); 6. Lead wall; 7. PVC gaiter; 8. Sodium
lamps; 9. Roof shielding

efficiency vacuum cleaner situated in the cell, is sometimes fitted around the wheel
to minimize spread of sample debris; cutting operations are observed through the cell
window with a stereomicroscope fitted with a long working distance objective. Materials which are cracked or porous i.e. metal clad UO_2 fuel section can be embedded in
cold curing epoxy resin to prevent their fragmentation during cutting. After cut off,
which is usually carried out in a separate cell, samples are transferred to a cell for mounting and mechanical polishing.

Cold curing resins are also used to embed samples in preformed mounts although
heated presses are sometimes used. Vacuum impregnation of cracked or porous samples

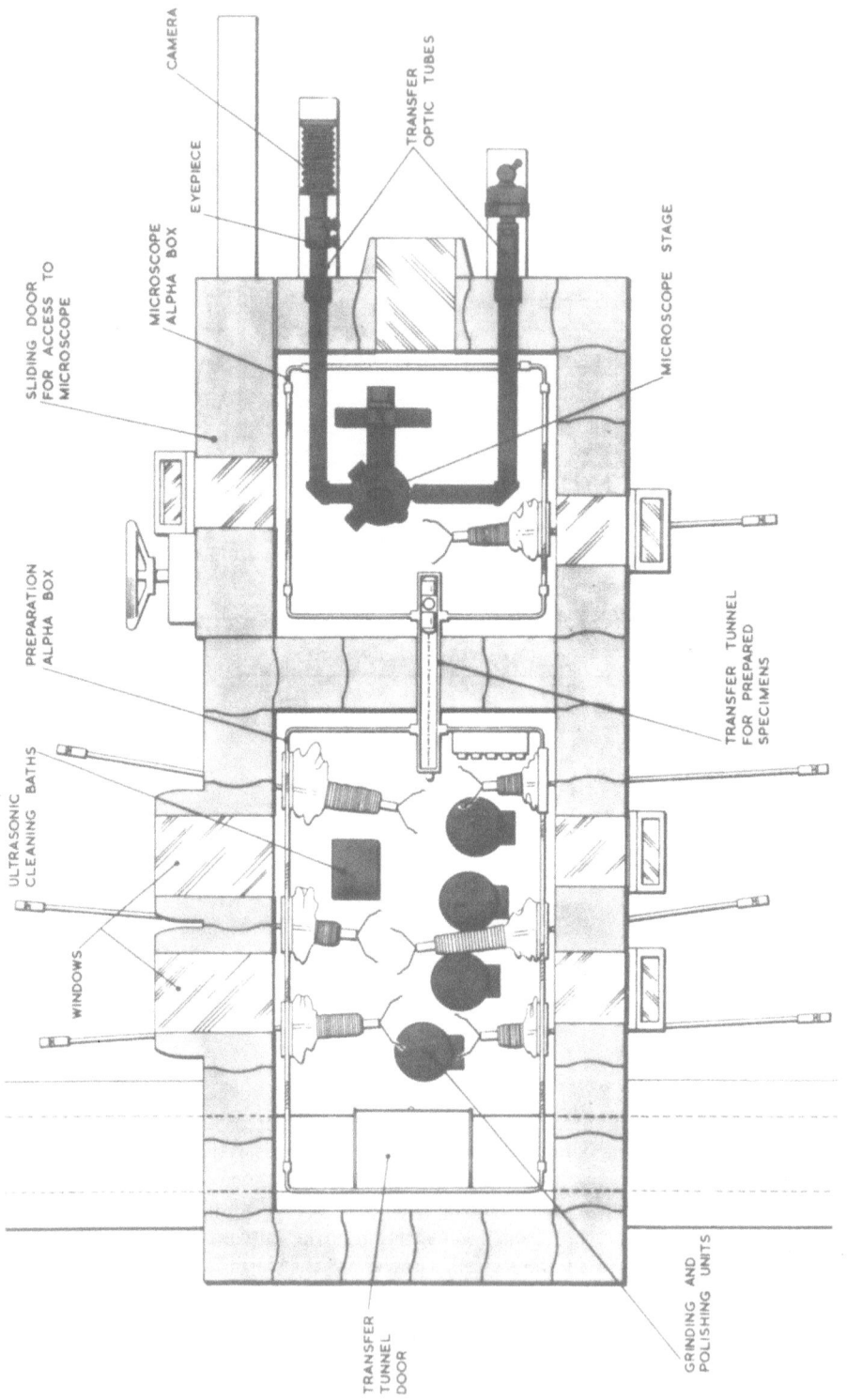

Fig. 3. Alpha/Gamma Metallographic Preparation and Examination Cell

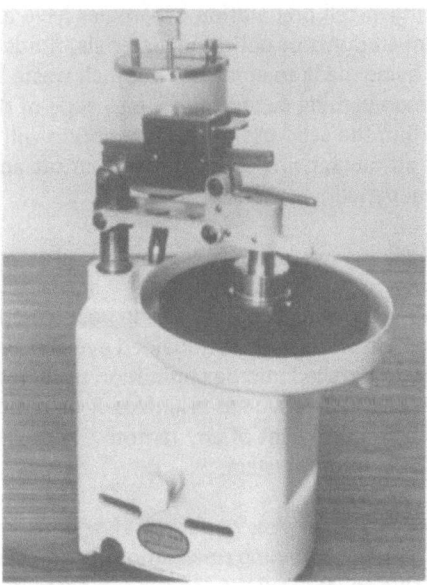

Fig. 4. Machine for Sample Preparation in Glove Box or Shielded Cell Operations

is carried out to eliminate staining during microscope examination. Grinding and polishing operations using conventional rotary and vibratory machines are usually performed in separate cells to prevent grit cross-contamination. Specially designed machines are also available and Fig. 4 shows a machine designed for use in glove boxes and shielded cells; a feature of the machine is that it can be dismantled into units for ease of transfer in and out of cells. Conventional abrasive materials such as silicon carbide papers and diamond pastes are used for grinding stages. The load applied to the sample and time of grinding are carefully controlled especially when preparing small samples, such as coated particle spheres 0.5 to 1 mm diameter used as fuel for high temperature reactors. Diamond pastes are also used for polishing stages of preparation using conventional lap materials charged with paste prior to posting into cell, further paste can be applied remotely.

Between each preparation step the sample is immersed in an ultrasonic bath. This treatment serves two purposes; to prevent the carry over of abrasive debris, and to remove surface contamination (this is most important prior to transferring samples to the microscope cell which should be kept as clean as possible to allow maintenance via gloves.)

Chemical and electrolytic techniques are employed for etching polished samples. Much skill is required to produce a satisfactory etch in conventional metallography; remote operators can produce comparable results although considerably more time and patience is required. Cathodic vacuum etching (3) is used in many hot laboratories and is especially useful where composite samples are being prepared. Liquids are not required and so the waste disposal problem is eased.

All materials used during in-cell preparation of samples have to be treated as radio-active waste and minimum amounts of polishing materials, fluids and etchants are used. Efforts are continuously being made to eliminate as much waste as possible and a current approach is to replace silicon carbide papers with a new type of diamond compound(4) which can be recycled within the cell; use of this compound will replace all grades of silicon carbide paper and although being developed for remote application it should be of interest to conventional metallographers.

OPTICAL EXAMINATION

Optical instruments are used at many stages during post irradiation examination. The process of optical examination starts with the unaided eye and progresses through low power macro examination to detailed microexamination using research microscopes capable of employing all usual techniques such as bright/dark field, polarized light, microhardness testing, etc. The prime requirement of any remote optical system used is that its performance equals that of a normal system.

However, radiation causes normal glass to darken after a time and to reduce this effect radiation resistant glasses containing small amounts of ceria have been developed for use in shielded cells. Due to its proximity to the radioactivity samples the most vulnerable component in an optical system is the objective. Fig. 5 shows the difference in behavior of two objectives, one with ordinary glass and the other with radiation-resistant glass. It is possible to use standard lenses and discard these when their performance deteriorates, and this procedure has been successfully used at Harwell for macro and micro optical systems. Up to two years use has been obtained with some microscope objectives.

MACRO EXAMINATION

In shielded cells, standard instruments can only be employed after modification to allow their use over extended distances. In most cases specially designed instruments are used. There are several methods of obtaining a magnified view of materials in a shielded cell.
(a) Binoculars, modified for close viewing or a stereomicroscope fitted with long working distance objectives (available for distances up to 2 metres) can be used for viewing through the cell window. These instruments are used to view operations such as cutting, mounting, etching etc.
(b) Periscopes fitted through the cell wall. There are several commercially available instruments which utilize the periscope principle. Large periscopes capable of scanning the entire cell working area, are fitted in concrete shielded cells for viewing general cell operations. Fig. 6 shows 3 methods of periscope viewing in lead cells.
(c) Macro instruments. These are usually used for examination of polished samples, in vertical incident illumination. Such an instrument is shown in Fig. 7 (5). The instrument is supported in the cell wall (this model is used in a lead shielded cell); internal shielding is arranged so that direct radiation 'shine' is eliminated. Magnifications up to approximately 20X can be provided.

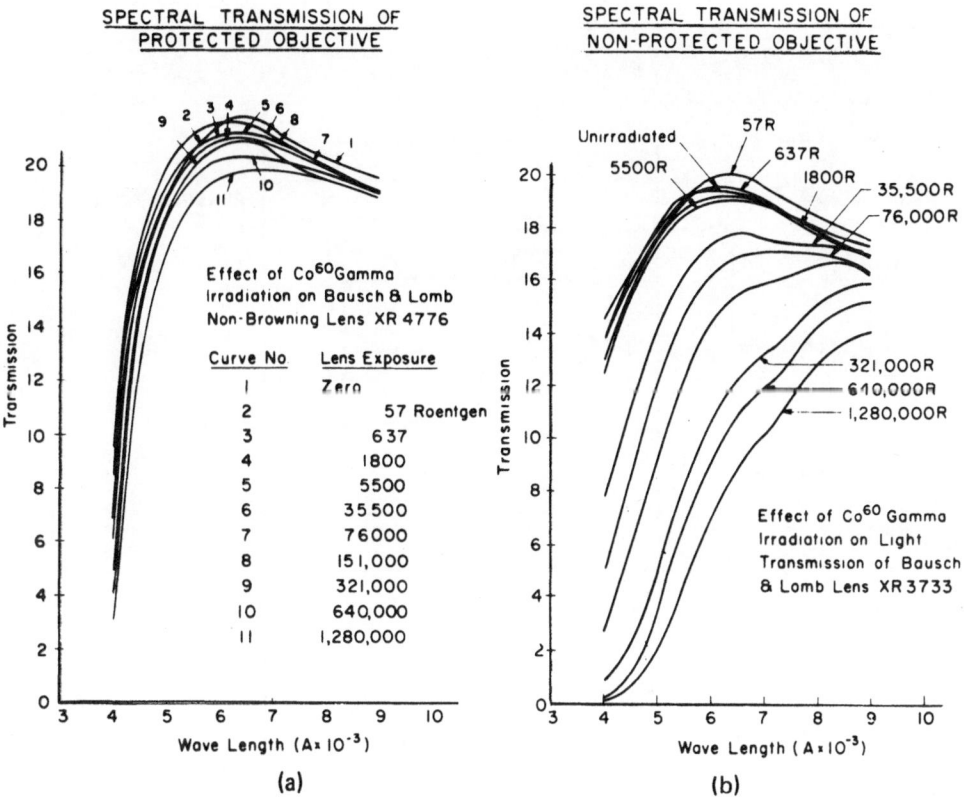

Fig. 5. Effect of Irradiation on Behavior of Objectives. (a) Constructed with Radiation-Resistant Glass, (b) Constructed with Conventional Glass.

A versatile macro camera has been designed at Harwell for macro examination of a wide range of sample materials(6). The camera shown in section (Fig.8) employs Zeiss (West Germany) luminar lenses supported on a sliding lens changer. One lens is used to photograph unprepared surfaces e.g. fuel element cans, irradiation assemblies in a wide range of illumination conditions provided by a ring flash unit; the lens diaphragm can be remotely operated and this ensures good control over depth-of-field, an essential feature for photographing objects such as tubes, spherical fuel particles, annular fuel bodies etc. A second lens, fitted with a vertical illuminator, is used for polished samples. Focussing is carried out by adjusting the stage height. The lenses are used within their design limits and results obtained with the camera are of the same quality as those obtained with a standard macro camera. The lens assembly and stage unit are shown in Fig. 9.

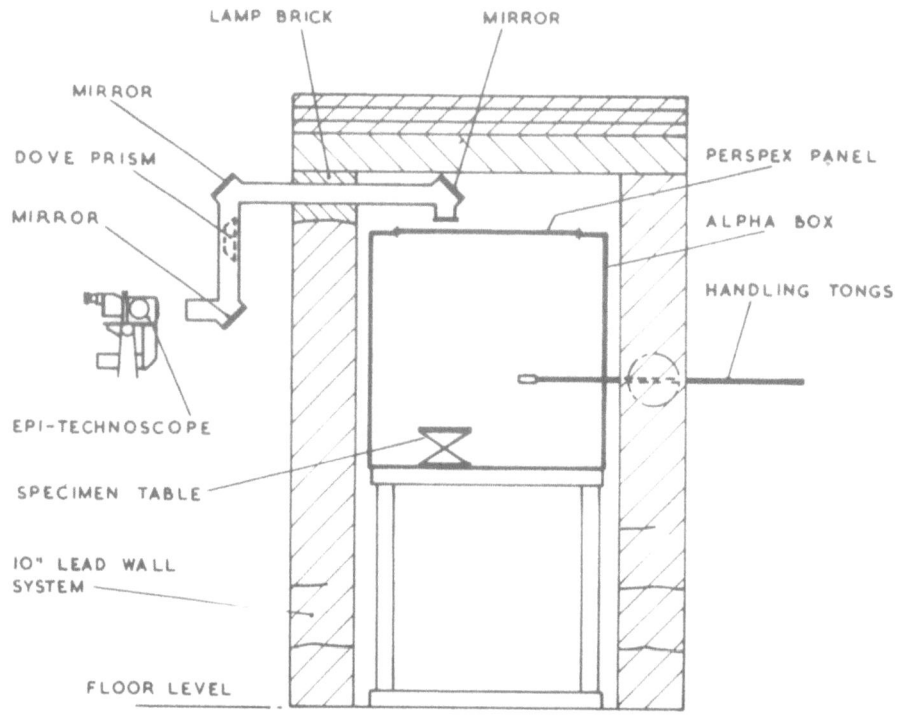

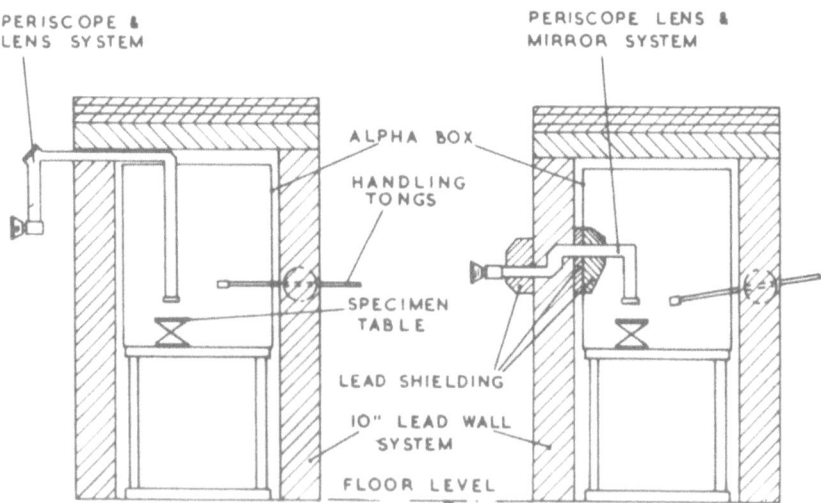

Fig. 6. Methods of Utilizing Periscope Principle

Fig. 7. Instrument for Macro Examination of Polished Samples

MICROSCOPE EXAMINATION

After macro-examination, samples are transferred to a shielded optical microscope for examination. There are basically two methods of providing a microscope for remote operation:
(a) Commercially available model ready for installation in a shielded cell, and
(b) Standard microscope modified by user.
In both cases the microscope should be fitted for all normal optical techniques.

As well as providing easy access to the microscope for maintenance, the following features are highly desirable in a remote microscope:

(a) All controls should be easily accessible at the viewing position.
(b) Objectives should be parfocal and mounted on a motorized changer.
(c) The image should be parfocal at eyepiece and camera systems with easy interchange.
(d) Large stage movement up to 50 mm x 50 mm.
(e) Optic transfer tubes should be as short as possible.
(f) Microhardness tester should be available (preferably on objective changer).

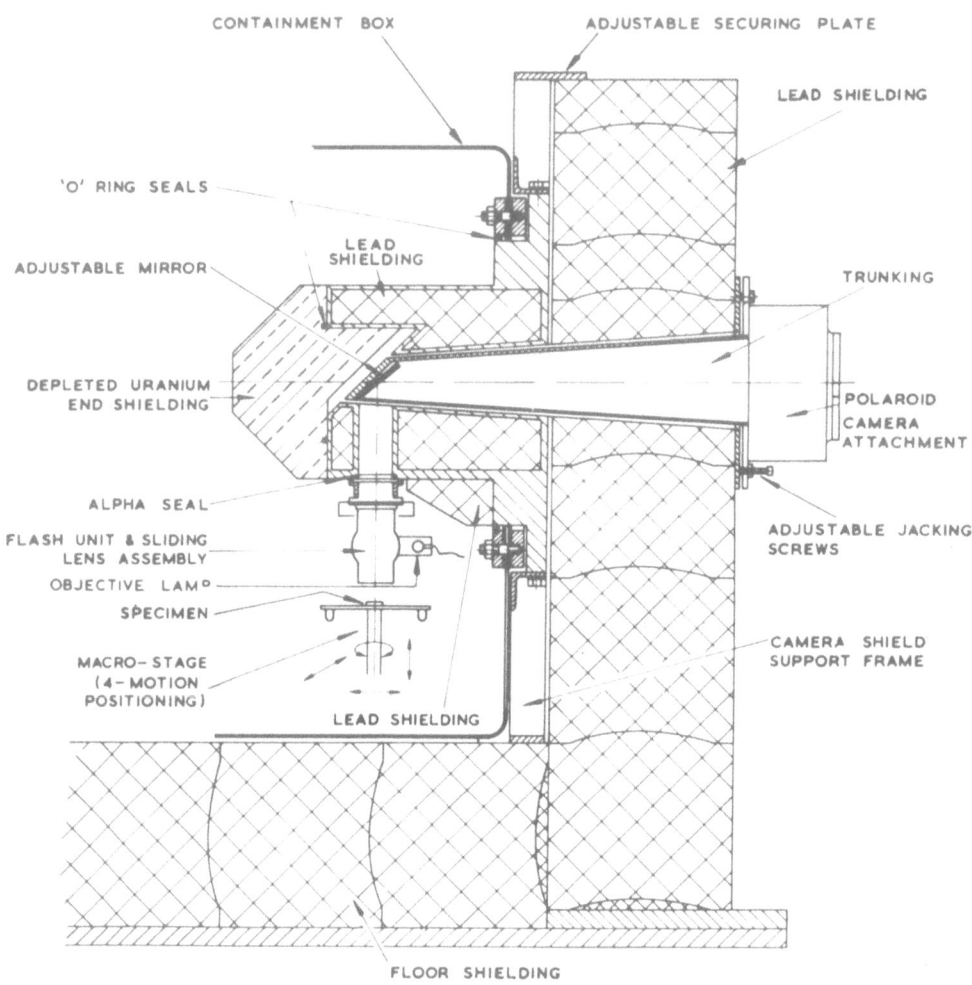

Fig. 8. Arrangement of Shielded Macro Camera

COMMERCIAL REMOTE MICROSCOPES

Several microscope manufacturers now offer a remote microscope for installation in a shielded cell. Each manufacturer has adopted a different approach and in this section brief details are given of some of the models available, emphasizing their advantages and disadvantages.

Early commercial designs of remote microscope equipment were in many cases modelled on standard instruments which had been modified at various atomic energy laboratories. At Harwell for example, a Reichert MeF model (7) had been modified for use in a lead cell and the first remote microscopes from the Reichert Company were

Fig. 9. Lens Assembly and Stage Unit

1. Luminar Lens for Unprepared Surfaces
2. Diaphragm Lever
3. Ringflash Unit
4. Epi-Luminar for Polished Samples
5. Lamp for Epi-Luminar
6. Uranium Shielding

based on this instrument. Details of the development of the Reichert remote microscope are described elsewhere (8). In the current Reichert remote microscope, known
as Telatom, the influence of the early instrument can still be seen but refinements have
been incorporated in the meantime. In this instrument the microscope controls are
operated predominantly through rotating mechanical connections which pass through
the cell wall; some microscope functions, such as stage drives and rotation, are operated
by miniature electric motors. The illumination and image are transferred by long optic
tubes cranked to prevent a direct radiation shine path. Figure 10 shows a Telatom microscope prior to installation in shielding. The mechanical and electrical controls can be
seen mounted at the viewing position. Figure 11 shows the mechanical connections for
operating the objective changer, diaphragms, polarizer/analyzer, phase contrast, micro
hardness tester. This design of microscope cannot be removed from the shielded cell
without major dismantling and all cleaning and maintenance operations have to be carried
out through gloves. Radiation resistant glass is used for all in-cell components and radiation resistant objectives are available.

Another design approach which resembles the Reichert instrument is used by the Union Optical Company of Japan. In their Farom model, all functions are operated by electric motors eliminating through-the-wall mechanical connections and thus easing installation. Cranked optic tubes are used which can be arranged to suit the customer's cell layout. A useful feature of this microscope is the method used to change objectives; the motorized objective changer is electrically linked to the stage lift which is also motorized. An objective is selected by depressing the appropriate button on the control, console, this raises the stage, rotates the changer until the desired objective is in place, and then lowers the stage. Radiation resistant glass is used for in-cell components. Removal of the microscope from the shielded cell is not possible.

A further improvement in the design of remote microscopes is incorporated in the MM5RT model manufactured by E. Leitz (Wetzlar, West Germany). All mechanical, electrical and optical connections between the microscope control/viewing position and the in-cell part of the microscope are routed through a single robust tunnel. Figure 12

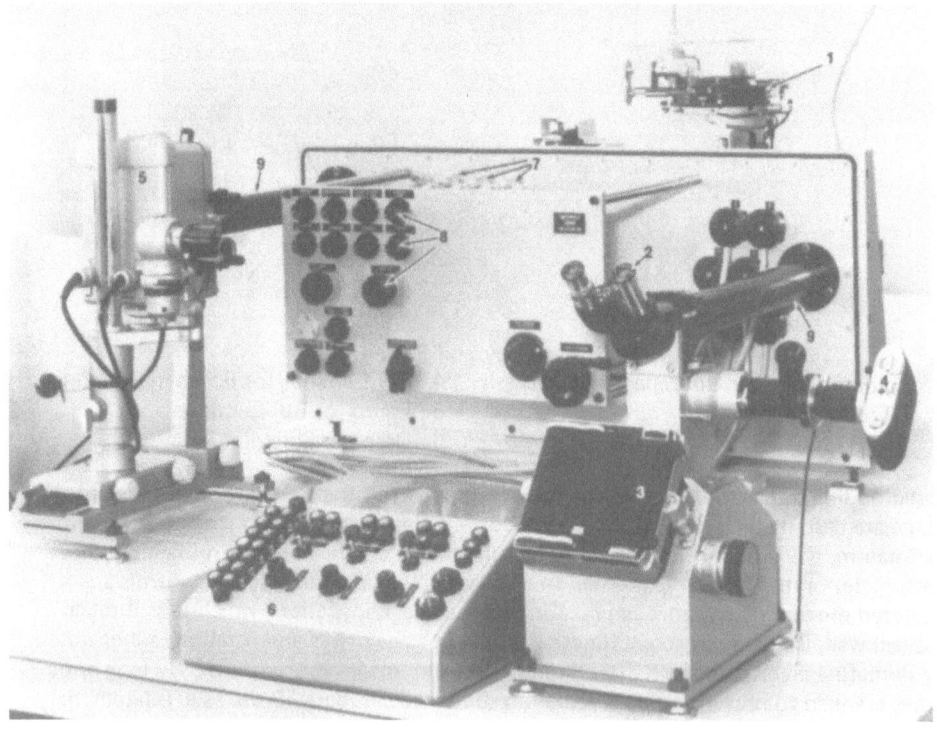

Fig. 10.   Telatom Microscope Prior to Installation

1. Microscope State
2. Binocular Viewer
3. 5" x 4" Camera
4. 35mm Camera
5. Lamp

6. Control Unit
7. Through-Wall Connectors
8. Rotary Controls
9. Optic Tubes

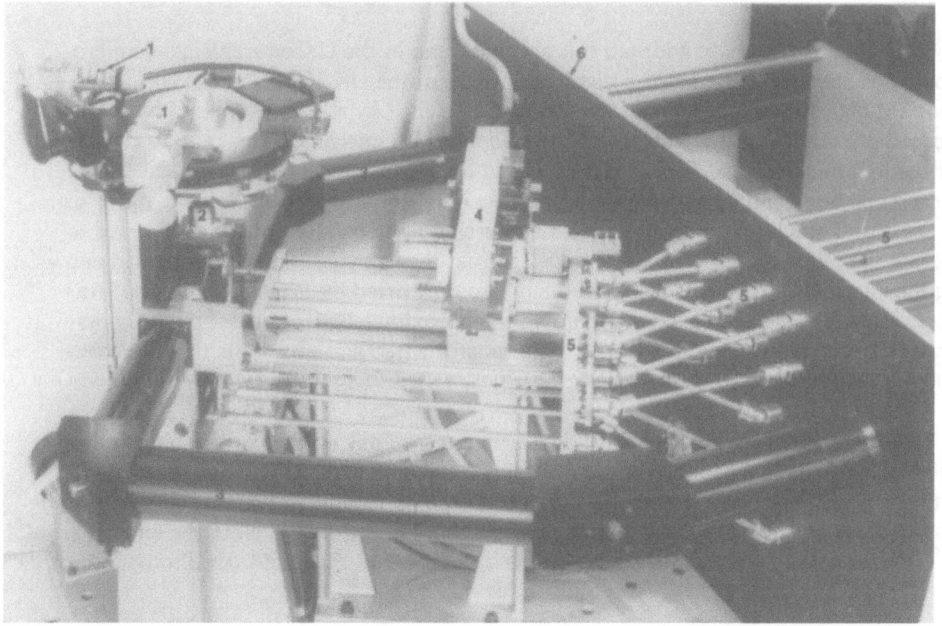

Fig. 11. Mechanical Connections for Operating Various Microscope Functions

1. Stage Motors   4. Objective Magazine
2. Objective   5. Through-Wall Connector
3. Optic Tubes   6. Alpha Box Panel

shows a side view of the instrument which is completely assembled and adjusted before delivery to the user. The integrated design permits the microscope to be mounted on a trolley allowing its withdrawal from the cell after removal of the front shielding. In-cell optics and objectives are made from radiation resistant glass. A microhardness tester can be mounted on the mechanically operated objective changer. A novel method of installation has been used at the Gulf Atomic Company Laboratories at San Diego, California (9) in which a Leitz remote microscope has been fitted within a concrete shielding wall obviating the need for a lead shielded cell.

Bausch and Lomb in the USA manufacture two remote microscopes:

(i) a model for installation in a blister (usually of cast iron) built onto the outside face of a concrete cell wall; the microscope stand is placed inside the blister and the lamp and viewer/camera are outside the cell.

(ii) an in-cell model for use in a concrete shielded cell; in this case the stand and lamp are sited inside the cell and the image is directed through the cell to the viewing/camera position outside the cell.

The blister model (10) is equipped for examination in bright field, polarized light and sensitive tint. All glass components inside the blister are made from radiation resistant glass. Due to its compact design and close proximity of the shielding, long cranked optic tubes are not necessary.

STANDARD MICROSCOPE MODIFIED BY THE USER

Many laboratories have adopted this approach and in the UK several Vickers microscopes have been modified at the Windscale works of the UKAEA. At Harwell several Neophot MkI microscopes manufactured by Carl Zeiss (Jena, East Germany) have been successfully used. Several features of the Neophot MkI microscope make it attractive for use in a shielded cell. Firstly it is mounted on a long optical bench with the lamp, microscope body and camera well spaced. Secondly, it has macro equipment which can be easily and quickly interchanged for the micro equipment. Remote controls used to operate the micro illuminator are arranged so that they can easily be disconnected allowing the micro unit to be removed and the macro unit fitted by means of a glove. The magnification range is thus extended down to approximately 10X; most remote microscopes can only be used with the micro illuminator fitted and this limits the minimum magnification usually to about 30X.

Stage controls are motorized with displacement read out on the control panel at the viewing position. A motorized triple objective changer is fitted which can be rotated with the stage lowered. A micro hardness tester can be fitted by means of a glove. Figure 13 illustrates how the modified microscope is arranged in the cell. Note the lamp is at 90° to the optical bench and the macro unit is not shown. Figure 14 is a photograph of

Fig. 12. Leitz Remote Microscope

1. Alpha Box Mounting Plate   5. Binocular Viewer
2. Microscope Stage   6. Viewing Screen
3. Tunnel   7. Camera
4. Lamps

the microscope in the cell before fitting the alpha box panel. A feature of the instrument is that, due to careful positioning of the microscope in the cell, long cranked optic tubes are unnecessary. Although in-cell glass components are made from normal glass, radiation darkening has not been a problem; one microscope has been in daily use for ten years with no significant change in its performance. Standard objectives are used and usually last for about two years before they are discarded.

A new metallographic facility is currently being built at Harwell which will be equipped with a Zeiss Neophot Mk II model (the Neophot Mk I model has been discontinued). This instrument is more compact than the previous model. Many of the features of the microscope have been retained in the remote version including the lamps, magnification variator and unique stage lift which is operated by simply moving a lever attached to the side of the base. All remote controls including focussing, objective changing, stage drives, etc. are operated by servo motors. The objective changer is fitted with five objectives and a microhardness tester. The microscope, mounted on a movable plinth, can be withdrawn from the shielding for major maintenance/repair if necessary. Once again due to careful positioning in the cell, special cranked optic tubes are not required. Figure 15 illustrates the modified microscope installed in a lead cell. All controls are grouped on either side of the viewing/camera unit. Figure 16 illustrates the new alpha/gamma metallography cell.

COMPARISON OF COMMERCIAL AND STANDARD MICROSCOPES

A remote microscope is a high cost item and much thought should be put in to select the most suitable instrument. The choice between the two methods of providing a microscope for remote operation will probably be made after consideration of exact user needs, market availability, timescale and cost. In the case of Commercial remote microscopes, some are easier to install in a shielded cell and this point is of prime importance when suitably skilled staff are not available. The use of a single route for all controls as well as easing installation, minimizes the risk of optical misalignment during use. In a standard microscope the instrument can be chosen to meet as nearly as possible the examination program requirements and cell layout. Since the microscope has to undergo considerable modification this approach should only be considered if suitable staff — at the user laboratory or specialist optical firm — are available.

Experience has shown that a modified standard microscope can be fitted in a shielded cell for about half the cost of a commercial remote microscope; running costs however, are higher due to radiation darkening of normal objectives.

ELECTRON—OPTICAL INSTRUMENTS

Electron microscopes are used routinely to examine irradiated materials. Replica electron microscopy is used to study details which are below the resolution limit of the optical microscope. The first few replicas taken of a sample in a shielded cell are usually heavily contaminated and so are discarded; subsequent replicas are low enough in activity to be transferred out of cell for treatment in a carbon evaporator/ shadowing unit. The final carbon replica is usually of low activity and can be safely transferred to an electron microscope, care is necessary however to prevent breakup of the replica which could become an ingestion hazard.

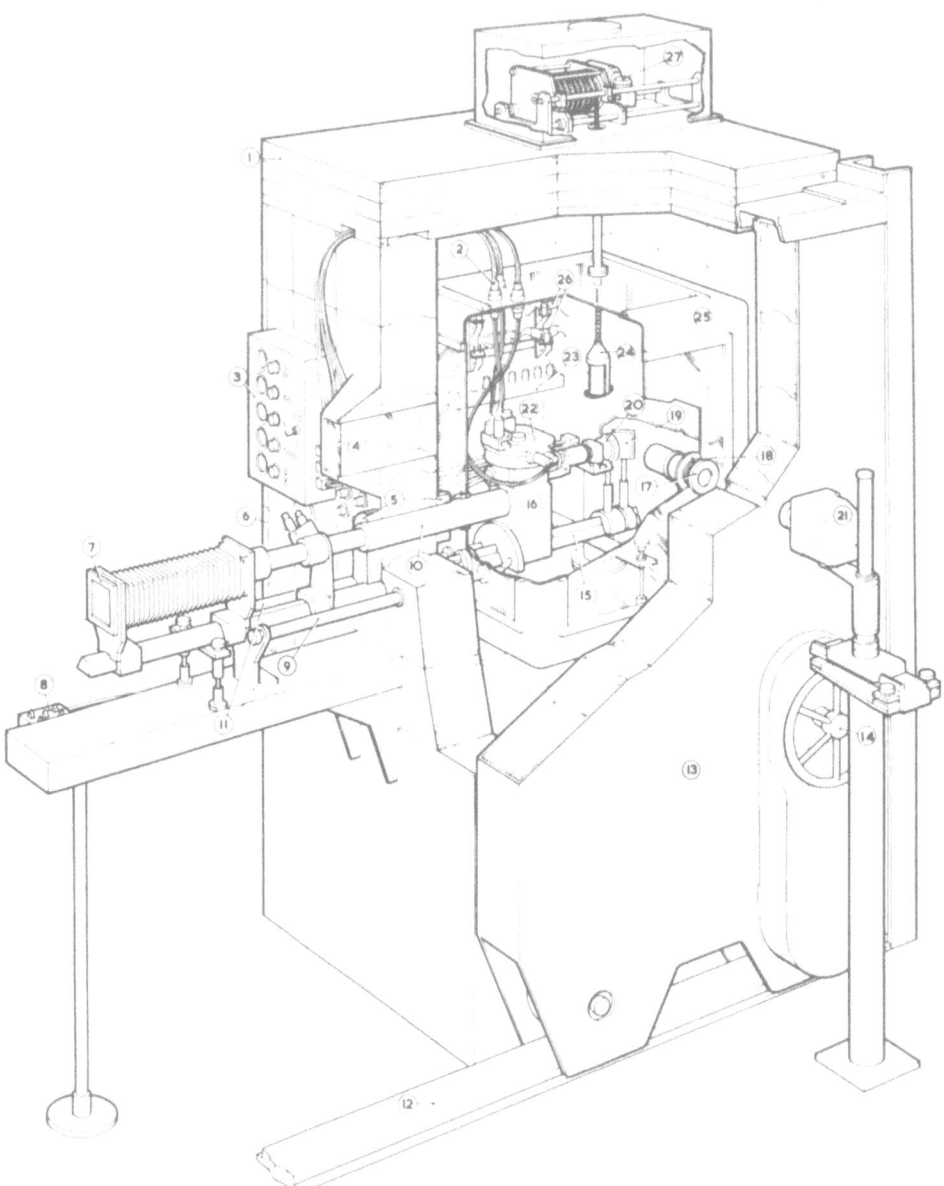

Fig. 13. Illustration of How Modified Microscope is Arranged in the Cell. (See Fig. 1 for Caption Detail).

Thin films of irradiated materials are also examined by transmission electron microscopy. Techniques have been developed for thinning bulk material (11), thin films are sufficiently low in activity to allow transfer to a conventional electron microscope, again care is necessary to avoid break up of the film. In both these techniques due to

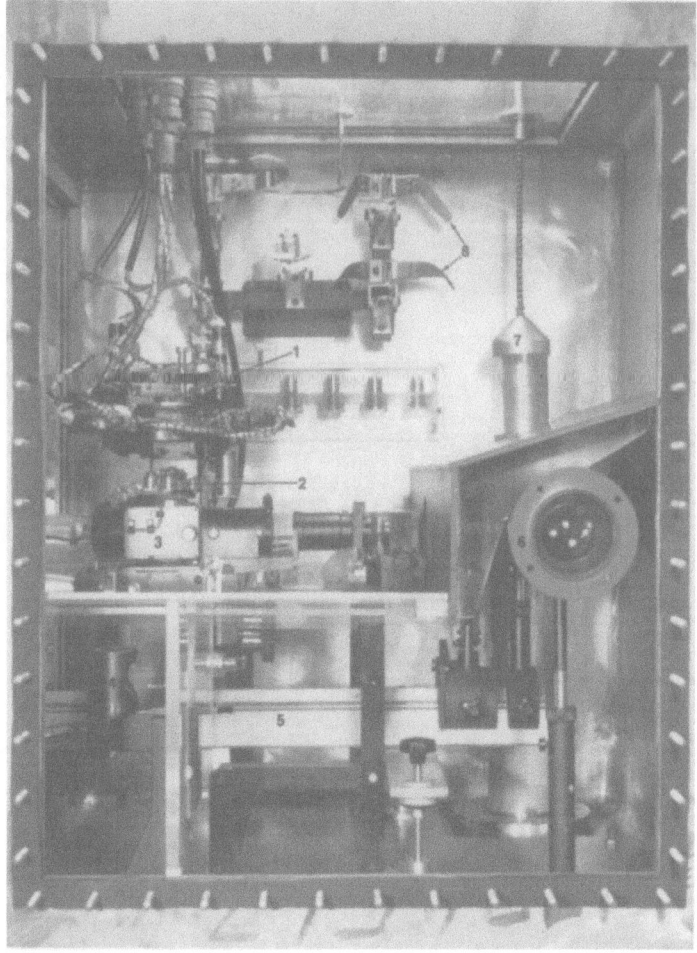

Fig. 14. Microscope in Cell Before Fitting Alpha Box Panel

1. Motorized Stage   5. Optical Bench
2. Motorized Objective Changer   6. Illumination Optic Tube
3. Micro Illuminator   7. Specimen Transfer Bucket
4. Macro Illuminator   8. Tong Fingers

the low activity levels of the replica/thin film, loading and unloading operations are not hazardous. In the scanning electron microscope (SEM) however, the sample can be considerably larger and therefore more active; in this case additional shielding is necessary to protect the operator during sample changing. At Harwell a Cambridge Instrument Co. 'Stereoscan' microscope has been modified for examining $\alpha\beta\gamma$ active samples (Fig. 17). The electron probe micro-analyzer is also employed for examining irradiated materials; at Harwell a Cambridge Instrument Co. 'Microscan V' has been modified for $\alpha\beta\gamma$ active samples (Fig. 18). A detailed description of these techniques is given by Lambert (12) and Rosenbaum (13).

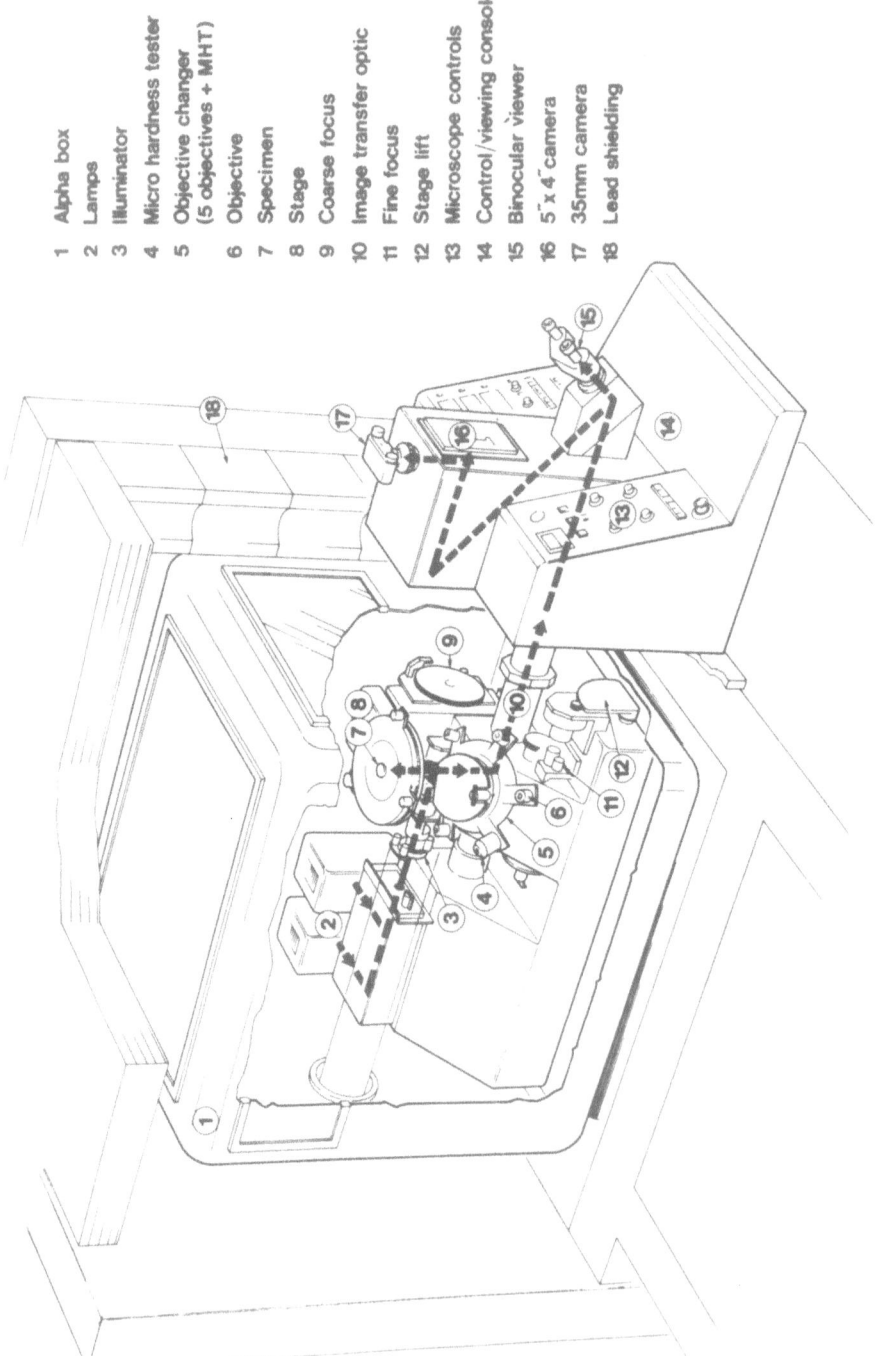

1 Alpha box
2 Lamps
3 Illuminator
4 Micro hardness tester
5 Objective changer
  (5 objectives + MHT)
6 Objective
7 Specimen
8 Stage
9 Coarse focus
10 Image transfer optic
11 Fine focus
12 Stage lift
13 Microscope controls
14 Control/viewing console
15 Binocular viewer
16 5″ x 4″ camera
17 35mm camera
18 Lead shielding

Fig. 15. Zeiss Neophot Microscope in shielded cell

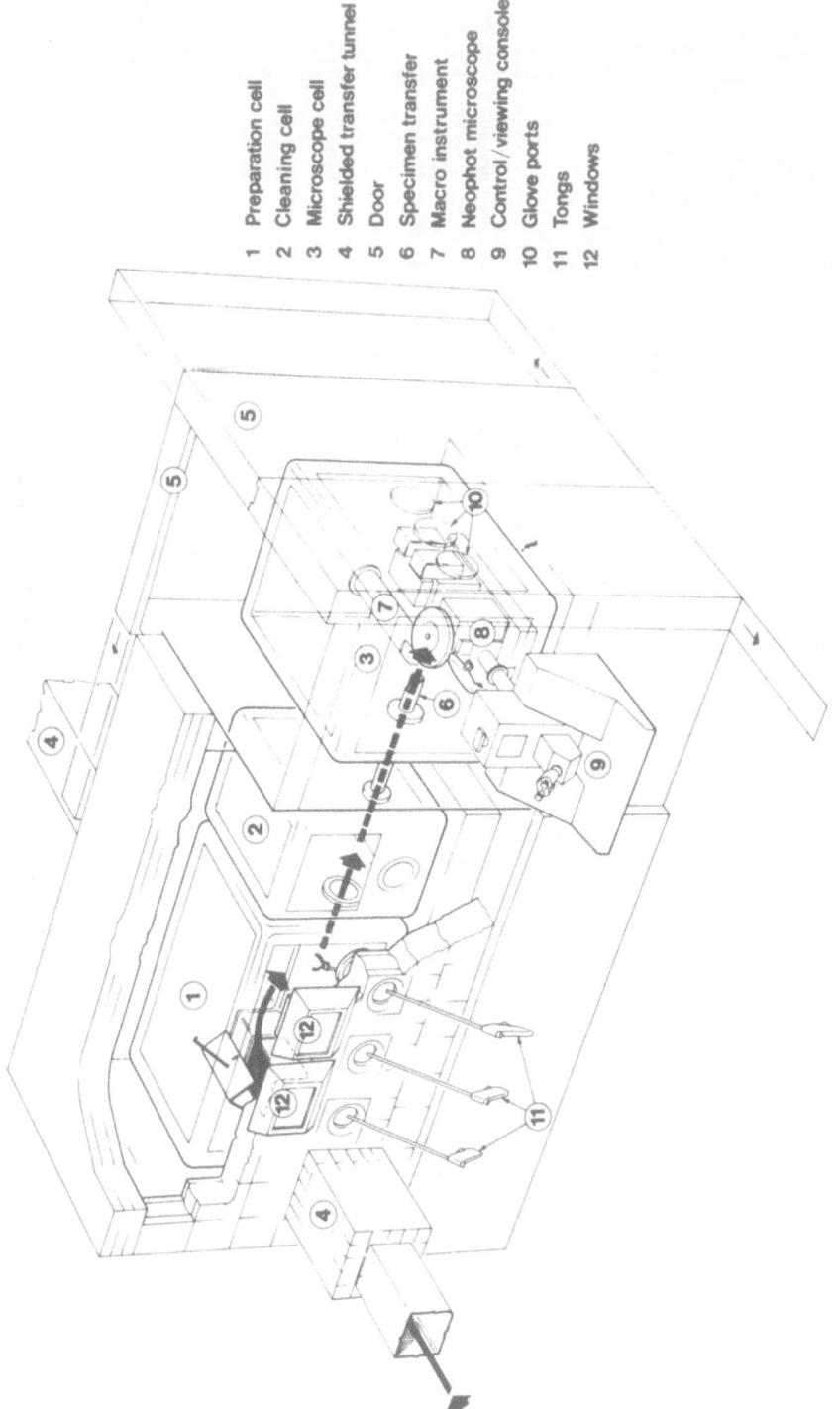

1 Preparation cell
2 Cleaning cell
3 Microscope cell
4 Shielded transfer tunnel
5 Door
6 Specimen transfer
7 Macro instrument
8 Neophot microscope
9 Control / viewing console
10 Glove ports
11 Tongs
12 Windows

Fig. 16. New Alpha/ Gamma Metallography Suite

ITEM	TITLE
1	Stereoscan Microscope
2	Specimen Stage
3	Specimen
4	Shielding Support Table
5	Mild Steel Shielding
6	Leaded Glass Window
7	Mini-Tong and Ball Unit
8	Box-Connector
9	Flexible Connection
10	Containment Box
11	Glove Port
12	Specimen Container Assembly
13	Container Mount Assembly
14	Side Viewing Leaded Glass Panel
15	Specimen Transfer Port

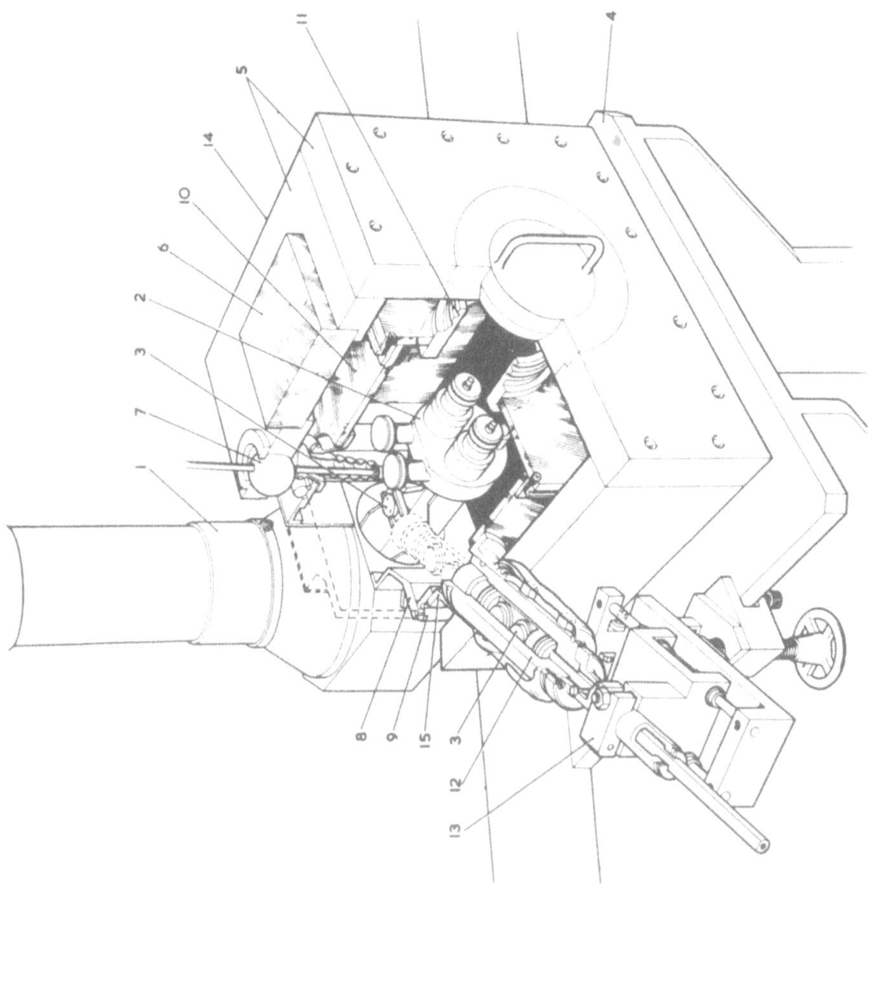

LEGEND

Fig. 17.   Shielded Stereoscan Microscope

LEGEND	
ITEM	TITLE
1	Microscan 5
2	Specimen Loading Cell
3	Extract Filter
4	'Dag' Applicator
5	Capsule Opening Tool
6	Capsule Clamping Tool
7	Transfer Hatch
8	Containment Box
9	Tool for Final Positioning of Specimen Stage
10	Transfer Tunnel
11	Transfer Tunnel Seal
12	Guide Channel for Specimen Stage
13	Modified Specimen Chamber Door and Vacuum Seal
14	Modified Specimen Stage
15	G.E.C. Heavy Alloy Specimen Holder
16	Specimen Chamber Shield on Swinging Arm
17	Sliding Shield for 9
18	Specimen Chamber Top Shield (Heavy Alloy)
19	Extended Filament Centering Rods
20	Modified Lever for Microscope Lens Positioning
21	Lead Glass Window
22	Lamp (two)
23	Tunnel Shield
24	Two Camera Positions

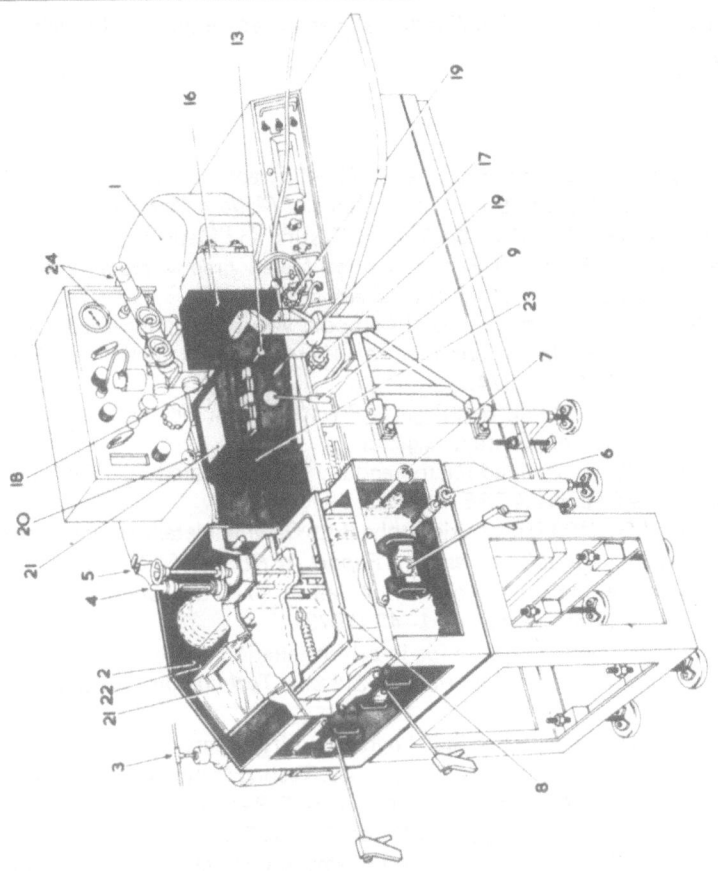

Fig. 18. Shielded Microscan 5

SUMMARY

Remote metallography is an important part of the post irradiation examination of nuclear reactor materials. Conventional metallographic preparation and examination techniques have been applied to irradiated samples.

This paper describes how radioactive samples, although hazardous, can be safely examined in shielded cells. Specially designed optical microscopes, as well as standard models modified by the user, have been introduced which enable all normal microscope techniques to be employed. Other instruments such as the electron microscope, for scanning and transmission, and electron probe microanalyzer have been adapted for use in this field.

Remote metallography has played a significant role in the successful development of nuclear reactor systems now extensively used for generation of electricity throughout the world.

ACKNOWLEDGMENTS

The author would like to thank colleagues at Harwell and elsewhere for help in preparing this paper.

REFERENCES

1. Gray, R.J., Long, E.L., Jr. and Richt, A.E., 'Metallography of Radioactive Materials at Oak Ridge National Laboratory', *Applications of Modern Metallographic Techniques,* ASTM STP480, American Society for Testing and Materials, pp 67–96 (1970).
2. Krautwedel, K.H., 'A Review of Remote Optical Metallography' *Metallography,* Volume 2, Numbers 2/3, pp 191–208 (September 1969).
3. Cain, F.M. Jr., 'Practical Applications of Cathodic Vacuum Etching' *Metallographic Specimen Preparation — Optical and Electron Microscopy,* Ed. McCall and Mueller, Plenum Press, pp 107–232 (1974).
4. Engis Ltd. Maidstone England. R.J. Swan — Private Communication.
5. Applied Optics, Couldsdon, Surrey, England.
6. Brown, P.E. et al, 'Lead Shielded Micro-handling cell for Coated Fuel Particles'. *Harwell Unclassified Report AERE R—6702. 1971.*
7. Greer–Spencer, J.G., '100 MeV Curie Alpha Beta Gamma Metallographic Suite — Engineering Design'. *Harwell Unclassified Report AERE R—2549. 1961.*
8. Evans, J.H., 'Remote Control Microscopy' *Advances in Optical and Electron Microscopy,* Volume 5, Academic Press, pp 1–42 (1973).
9. Cochran, F.L. et al, 'Design Criteria for In-cell Installation of Metallograph stages' *Proceedings of 19th Conference on Remote Systems Technology,* American Nuclear Society, 1971.
10. Gray, R.J. ' The Present Status of Metallography' Fifty Years of Progress in Metallographic Techniques, *ASTM STP 430,* Am. Soc. Testing Materials, pp 17–62 (1968).
11. Bainbridge, J.E., *Harwell Unclassified Report AERE R—5677,* 1968.
12. Lambert, J.D.B., *Nuclear Energy,* Jan, March, May 1968.
13. Rosenbaum, H.S. 'On the Use of Electron Beams to Characterize the Microstructure of Radioactive Materials', *Metallographic Specimen Preparation. Optical and Electron Microscopy.* Plenum Press, 1974.

HOLOGRAPHIC MICROSCOPY

MARY E. COX *

INTRODUCTION

The goal in the design of a holographic microscope is to record and reconstruct a three-dimensional image of a microscopic object so as to preserve both amplitude and phase information. Using standard microscopic techniques where possible, the holographic microscope is a tool to be used by any scientist requiring this amplitude and phase information. This paper presents a discussion of state-of-the-art in holographic microscopy, including current techniques, basic principles of holography especially as they apply to holographic microscopy, and successful applications. The goal of this paper is to aid the reader in making a decision about the utility of holographic microscopy in a particular research situation.

A few words of caution are required at this point. No commercial, off-the-shelf holographic microscopes are currently available. Each application of holographic microscopy requires detailed evaluation of the parameters involved. Every holographic microscope must also be custom tailored to the application desired. This requires close cooperation between the scientist with the application and the designer of the holographic microscope. The scientist must be able to specify in detail just how the holographic microscope is to be used. Likewise, the designer must work very closely with the scientist to see that these specifications are incorporated into the holographic microscope, and then must instruct the scientist in the proper use of the holographic microscope. With the closest cooperation between scientist and designer, a useful tool can be created. Only with this close cooperation can the holographic microscope serve the scientist in the manner desired.

HOLOGRAPHY

Dennis Gabor introduced holography as a "new two-step method of optical imagery"(1). In attempting to improve the resolution of the electron microscope, Gabor investigated the two-step imaging process used by Bragg's X-ray microscope(2). The basic concept in Bragg's method is a double-diffraction process, which is the crux of the holographic process. The distribution of light diffracted by an object can be represented as a Fourier transform of the light at the object. If this light distribution is recorded photographically, a second diffraction of light from this photographic plate is a Fourier transform of the Fourier transform of the object, which is an image of the object itself.

*Department of Physics & Astronomy, University of Michigan—Flint, Flint, MI 48503

Gabor realized that if the object was to be accurately reconstructed, the amplitudes and phases of all waves must be preserved. He introduced a standard reference wave to be added to the diffracted object wave prior to recording the hologram. This reference wave was more intense than the wave diffracted by the object, so that the amplitude of the reference wave was modulated by the object wave. A processed recording of the resulting modulated diffracted wave constitutes the hologram. Irradiation of the processed hologram results in a second diffraction, resulting in the reconstruction of the original object wave field.

Gabor's original work in 1948(3) was unsuccessful in practice, because of various instabilities in his system. The principle of adding a coherent reference wave to an object wave was sound. Aside from a few brief notes in optics textbooks, the entire subject of holography lay dormant until the early 1960's.

Leith and Upatnieks(4) demonstrated a successful method of recording and reconstructing holograms using an off-axis reference wave. This process is similar to placing a signal wave on a carrier wave, producing sidebands. The two reconstructed wave fields are then separated in space, and the hologram can be analyzed in a manner similar to a diffraction grating or a zone plate. Later, Leith and Upatnieks introduced illumination of objects(5).

Holograms can now be recorded of almost any object that can reflect or transmit light. The reconstructions from diffusely illuminated objects are pleasing to the eye and are truly three-dimensional in character, complete with parallax and large depth of focus.

BASIC PRINCIPLES OF HOLOGRAPHY

Fig. 1 shows a generalized holographic recording geometry. The object field may be produced by reflection, transmission, or diffraction, or any combination of these. The reference field may be produced in a wide variety of ways, but must be reproducible. The interference pattern produced by these two fields is recorded as an intensity pattern, often on photographic film. When properly processed, this recorded intensity pattern is called a hologram.

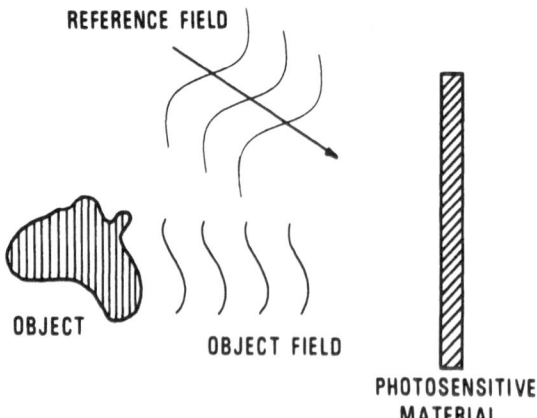

Fig. 1. Generalized holographic recording geometry

The object field may be reconstructed as shown in Fig. 2. In this figure the reconstructing field is an exact duplicate of the original reference field. Two jugate image fields will be produced, one identical with the original object, but located on the "virtual image" side of the hologram, and the other on the "real image" side of the hologram. To be seen, the "virtual image" requires auxiliary optics, such as the human eye. The "real image" can be projected directly on a screen.

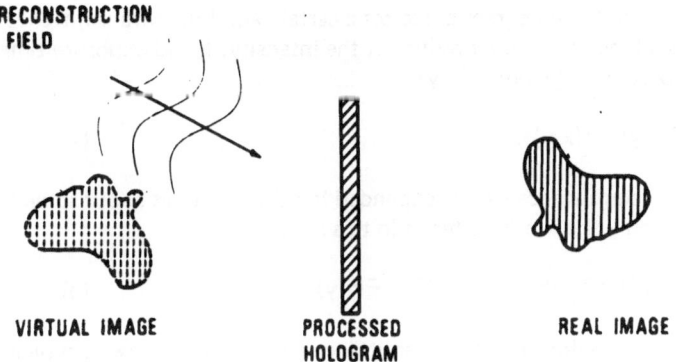

RECONSTRUCTION
FIELD

VIRTUAL IMAGE   PROCESSED   REAL IMAGE
                       HOLOGRAM

Fig. 2. Generalized holographic reconstruction geometry

To understand the production of the twin images, let us represent the original object wave at the surface of the recording medium(6).

$$u_o(x,y) = a(x,y)e^{-i\Theta(x,y)} \tag{1}$$

where $a(x,y)$ represents the amplitude and $\Theta(x,y)$ the phase at the point (x,y). The reference wave at the surface of the recording medium is

$$u_R(x,y) = A(x,y)e^{-i\Psi(x,y)} \tag{2}$$

where $A(x,y)$ represents the amplitude and $\Psi(x,y)$ the phase at the point (x,y). Since only intensity variations can be recorded, the intensity at the point (x,y) is

$$I(x,y) = |u_R(x,y) + u_o(x,y)|^2. \tag{3}$$

The absolute value of a complex quantity is real and given by

$$|p(x,y)| \overset{\Delta}{=} |p(x,y)\ p^*(x,y)|^{1/2}. \tag{4}$$

Thus,

$$I(x,y) = |u_R|^2 + |u_o|^2 + u_R^* u_o + u_R u_o^*. \tag{5}$$

or,

$$I(x,y) = |A(x,y)|^2 + |a(x,y)|^2$$

$$+ 2A(x,y)a(x,y)\cos|\Psi(x,y) - \Theta(x,y)| . \qquad (6)$$

The intensity at the point (x,y) is the sum of three terms. The first two give information on the intensities of the reference and object waves, respectively. The third term depends on the interference between the object and reference wave. When this intensity pattern has been recorded and processed, the result is a hologram.

After processing the hologram possesses a certain amplitude transmittance at the point (x,y) which is a function of the product of the intensity, I, and exposure time, t. Using E(x,y) for exposure at the point (x,y).

$$E(x,y) = I(x,y)t, \qquad (7)$$

the amplitude transmittance can be expanded in a Taylor series about some bias exposure, E_B. Retaining only the first two terms in this series

$$T_A(E) = T_A(E_B) + T_A'(E) \cdot E(x,y), \qquad (8)$$

where $T_A'(E)$ is the slope of the transmittance curve. Fig. 3 shows a typical transmittance curve for a suitable recording medium used for holography.

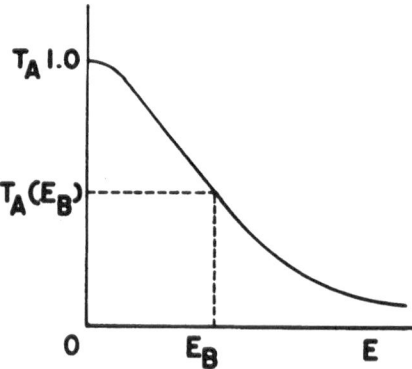

Fig. 3. Amplitude transmittance vs. exposure curve for typical recording medium

To reconstruct the object fields, the processed hologram is illuminated by a reconstruction wave, $u_r(x,y)$. The light transmitted by the hologram is

$$u_r(x,y)T_A(E) = T_A(E_B)u_r(x,y) + T_A'(E)u_r(x,y)E(x,y). \qquad (9)$$

Thus,

$$u_r(x,y)T_A(E) = T_A(E_B)u_r(x,y)$$

$$+T_A{}'(E)u_r(x,y) \cdot t \cdot [\, |u_R|^2 + |u_o|^2 + u_R{}^*u_o + u_Ru_o{}^*].\quad (10)$$

The first term represents the attenuated reconstruction wave. The terms proportional to $|u_r|^2$ and $|u_o|^2$ represent attenuated reconstruction wave which would occur if the hologram had been exposed to only the reference or object waves, respectively.

If the object field subtends a large solid angle at the point (x,y) it is possible for the $|u_o|^2$ terms to contain interference effects. That is, two object points can produce waves which interfere at the hologram plane. These intermodulation terms give rise to a spreading of the reconstruction beam and some object noise in the reconstructed image. We will discuss noise in detail later in this article. The terms of critical interest in image reconstruction are those proportional to $u_R{}^*\, u_o$ and $u_Ru_o{}^*$.

If the reconstruction wave is identical to the reference wave, that is, if

$$u_r(x,y) \equiv u_R(x,y) \qquad\qquad\qquad\qquad (11)$$

then

$$T_A{}'(E)tu_r(x,y)u_R{}^*(x,y)u_o(x,y)$$

$$= T_A{}'(E)t\,|u_R|^2\, u_o{}^*\,(x,y). \qquad\qquad (12)$$

This is a multiplicative constant times the original wavefront from the object. The exposing and processing have altered the original object field by some multiplicative constant which affects the overall intensity of the image. The original object field is reproduced in both amplitude and phase variations.

The other term becomes, with reference field equal to the conjugate of the reconstruction field,

$$T_A{}'(E)tu_r(x,y)u_r(x,y)u_o{}^*\,(x,y)$$

$$=T_A{}'(E)t\,|u_R|^2 u_o{}^*\,(x,y). \qquad\qquad (13)$$

This term produces the complex conjugate of the original object. Again the multiplicative constants affect overall intensity, but do not affect amplitude and phase variations as they occurred on the original object.

Originally, the production of the two images was detrimental since the two fields overlapped in space. However, when Leith and Upatnieks added the reference wave as an off-axis carrier wave, the twin images could be separated in space and studied separately. Now one routinely designs a holographic system which will enhance one image at the expense of the other.

RECORDING MEDIUM

By far the most popular medium used to record a hologram is photographic film. It is usually inexpensive, readily available, easily handled, and convenient. Most applications of holography are easily adaptable to using photographic film as the recording medium. For any particular application choosing the proper film is as important as any other parameter in the system. It is best to consult with the film manufacturers at this point to assure the correct choice(7).

Many holographers use photochromics, photopolymers, thermoplastics, dichromated gelatin, and electrooptic crystals. Each of these materials has special advantages in certain applications. Photochromics are erasable, and may be used over and over. Photopolymers, because of their rapid response time, can be used in situations where near real-time reconstruction is needed. Thermoplastics have high sensitivity with low noise levels, while dichromated gelatin produces holograms with unusually high efficiency, especially in the blue region of the visible spectrum. Electrooptic crystals produce high efficiency holograms and can be erased for reuse.

Let us examine the properties of any recording medium used for holography. First, of course, the materials must be sensitive to the wavelength used to record the hologram. Second, the material must be capable of recording the spatial intensity variations that occur when the object and reference waves interfere. For most applications, this means the medium must have a spatial frequency response above 1000 lines per millimeter. Third, the medium must be sensitive enough to record the intensities incident upon it. That is, the bias exposure must lie in a sensitive region of the medium's response curve. The transfer curve most useful for photographic film is the amplitude transmittance vs. exposure curve. Fig. 3 shows such a curve. It is obtained, usually, from the standard density vs. log exposure curve supplied by a manufacturer. For phase media, where intensity variations result in index of refraction changes, a very different transfer curve is needed. Lastly, the medium must be able to be processed in a manner that yields a good quality hologram(8). For our purposes, let us say that processing is critical, and must be carefully handled.

NOISE

Noise in the reconstructed image from a hologram arises from several sources. Film grain noise, object beam modulation noice, out-of-focus image noise, and speckle noise all contribute to reduce the signal-to-noise ratio in a holographic system. We wish to examine each source of noise to better understand how to minimize the noise in the reconstructed image.

An unexposed, processed piece of photographic film is not completely transparent. Some silver grains have been developed and scatter any incident illumination. All exposed and processed film possess some grain noise. This appears in the reconstructed image as a random illumination which may obscure fine details. In the case of reconstruction with coherent illumination, this random illumination may have regions of constructive interference, when the grain noise contributes illumination at a point greater than the object point illumination. If destructive interference occurs between the grain noise illumination and the object reconstruction, all information may be eliminated.

Object beam modulation can be observed by illuminating a moderately rough surface with coherent illumination. Viewed with the eye, the surface appears to have bright and dark regions. As the eye moves laterally, these bright and dark regions change location. The interference between waves originating on different object points is recorded in the hologram. Upon reconstruction, these bright and dark regions appear in the image. Again, information may be lost about some specific object features.

This noise is especially critical when viewing phase objects in transmission. In a holographic microscope when one is looking for small phase changes in an object which also introduces amplitude changes, details can be enhanced or eliminated.

Because of the three-dimensional nature of the reconstructed holographic image, out-of-focus image points contribute a non-uniform background. In even simple reconstructed images of widely scattered point objects, each out-of-focus image point appears as a series of concentric bright and dark rings. These rings may obscure image points that are in focus. They may also make location of the images difficult.

The most serious source of noise in the reconstructed image is speckle. Diffusers were originally introduced in holographic systems to reduce noise effects caused by dust particles and/or lens imperfections. In addition, the diffuser also increased the field of view of the hologram, sharpened the depth of focus of the three-dimensional image, and reduced the dynamic range requirement on the recording film. It was also observed that the diffuser introduced a granular noise pattern superimposed on the image(9). This noise pattern is referred to as speckle noise. The size of the speckle in the reconstructed image is determined by the size of the hologram used to reconstruct the image.

Speckle may be reduced by controlling the temporal and/or spatial coherence of the object beam. For holographic microscopy reducing the temporal coherence by using a broad band source has proven useful(10). Lowering the spatial coherence by using rotating diffusers or increasing the size of the source that is illuminating the object are not desirable. Both techniques reduce the system resolution. Ensemble averaging techniques can be used if the object is static, or slow moving(11). This technique creates superpositions of redundant low spatial frequency images in a wide bandwidth recording system to average speckle noise.

To minimize all sources of noise in a coherent holographic microscope, choose a recording medium whose achievable performances criterion is the system resolution limitation; then minimize recording medium nonlinearities by careful control of any processign required. Use good, clean optics. Finally, reduce speckle noise by minimizing the coherence of the source radiation in a manner compatible with the other system parameters.

DESIGNING A HOLOGRAPHIC MICROSCOPE

Individual design criteria will affect specific features of each holographic microscope. However, there are a number of general features that must be incorporated into the design of every holographic microscope. These general features are depth of field, field of view, resolution, magnification and aberration balancing. An examination of each of these features will show how they affect the design and operation of the holographic microscope.

All imaging systems, holographic or conventional, have a fundamental limit relating to focus properties and focal numbers. The focal number, F, is the ratio of the focal distance to the diameter of the aperture. The focus is distinct for lateral dimensions greater than a minimum length, L, where

$$L \sim F^{-1}$$

The focus in the depth dimension is distinct only for a depth less than a depth of field, D, where

$$D \sim F^2.$$

This means that if high resolution, low L, is to be achieved, then very small depth of focus, small D, will result. At any one time there can only be a region of about L x L x D dimensions in clear focus.

The holographic recording medium must be able to resolve sufficient information so that the object wave may be reconstructed. Just how much resolution is required depends on the object size, type of hologram, and geometric arrangement. In general, an in-line hologram requires the least amount of bandwidth and an off-axis diffuse-subject hologram the most. An off-axis hologram image suffers only a loss of field for limited recording medium resolution.

Another factor that affects the resolution in a holographic image is the size and bandwidth of the source or sources providing the reference and object beams. The minimum resolvable image point can be no less than the extent of the reference source. If the reference wave is ideally plane or spherical, but the illuminating wave is derived from other than a point source, then the point object will image as the illuminating source distribution. Thus, an ideal light source for recording a hologram would be a point source. Since sharpness is always diffraction limited, the light source need be no smaller than the size dictated by diffraction effects. Also, the bandwidth of the light source need not be zero, but only small enough so that any spreading of light in the image caused by finite bandwidth will be small compared with the spreading due to diffraction.

Aberrations enter holographic microscopy in the conventional, geometric sense and in the auxiliary optics used. It is a very complicated subject which is well-treated in the literature(12). All primary, third-order aberrations will be eliminated when plane waves are used for both the reference and object beams, when the offset angles are equal and on opposite sides of the normal to the hologram.

Magnification may be achieved in either of two ways. In the first, good quality optics are used to premagnify the object prior to recording the hologram(13). A conventional microscope produces a magnified real image beyond the eyepiece. This image can be the object for a hologram. In the second, magnification is achieved using poor quality optics between the object and the hologram(14). In this method, the reconstructed image from the hologram is played back through the poor quality optics. This eliminates the aberrations introduced by the optics, but the reconstructed real image must be examined through a conventional microscope.

In any given design, other parameters arise during the planning stage. Choice of wavelength is important, especially when highly colored objects are used. Highlighting of surface features can be achieved by using contrasting colors. The polarization state of the illuminating light can be of importance if the object is birefringent or optically active. Using a pulsed light source can be important if very short-lived, transient events are to be recorded. The nature of the environment surrounding the object can be important, especially if it is turbulent or has a different index of refraction. Optical path length changes can affect hologram quality and, thus, the quality of the reconstructed image.

APPLICATIONS OF HOLOGRAPHIC MICROSCOPY

Holographic microscopy has been successfully used in crystal studies, *in vivo* microcirculation studies, stained specimen studies, and polymer studies. Each application used a slightly different configuration of holographic microscope. Each shows the versatility of holographic microscopy, but also stresses the difficulties in achieving ideal imaging situations.

McFee(15) developed a holographic microscope for use in studying crystal growth from the melt. Figs. 4 and 5 show schematic diagrams of his experimental arrangement. McFee wished to study the crystal-melt interface. The object beam and reference beam are plane waves. The reference beam completely bypasses the crystal-growing furnace. The object beam is diffracted, refracted and scattered as it passes through the melt. The scattered light from the crystal-melt interface is collected by a silica lens inserted with its flat end into the melt. This silica lens, with a field lens, directs the object beam onto the photographic film. After processing, the hologram is inserted into a reconstruction geometry, where the silica and field lenses can be positioned to minimize aberrations. The real image is viewed with a conventional microscope.

This technique has proved useful when the object to be studied is in a foreign environment. Bubble dynamics in a living hamster were examined with a holographic microscope (16). The hamster was located in a hyperbaric chamber, where compression and decompression to stimulate dives could be accomplished. Various gases could be used. The capillaries in the cheek pouch are easily accessible for studying microcirculation hemodynamics. Fig. 6 shows the holographic microscope devised for this application. The object beam enters the side of the chamber, is reflected by $90°$, then is incident upon the cheek pouch. A large diameter lens is mounted on the inside of the chamber window, behind the microscope stage holding the hamster, to collect the scattered light and direct it onto the film plane. The reference beam is expanded and collimated, then passes around the chamber and onto the film. The lens and chamber window are reinserted in the reconstruction process. Again, the reconstructed real image is studied with a conventional microscope.

To reduce the numerical aperture of the diffracted object field at the film plan, van-Ligten and Osterberg place a set of well-corrected microscope optics between the object and the hologram(17). Fig. 7 shows this arrangement. The numerical aperture of the microscope objective must be very high, if fine detail is to be observed. However, at the film plane where the hologram is to be recorded, the numerical aperture of the object field has been reduced by an amount inversely proportional to the magnification. The reference beam is carried around the microscope. In the reconstruction one needs little, if any, additional magnification. This technique severely limits the field of view.

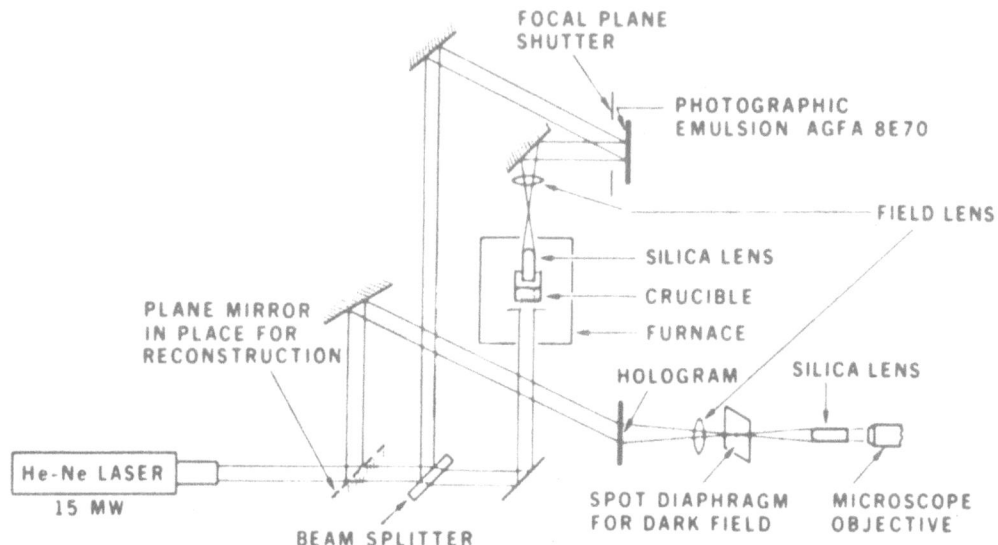

Fig. 4. Schematic diagram of holographic microscope developed by McFee (Ref. 15)

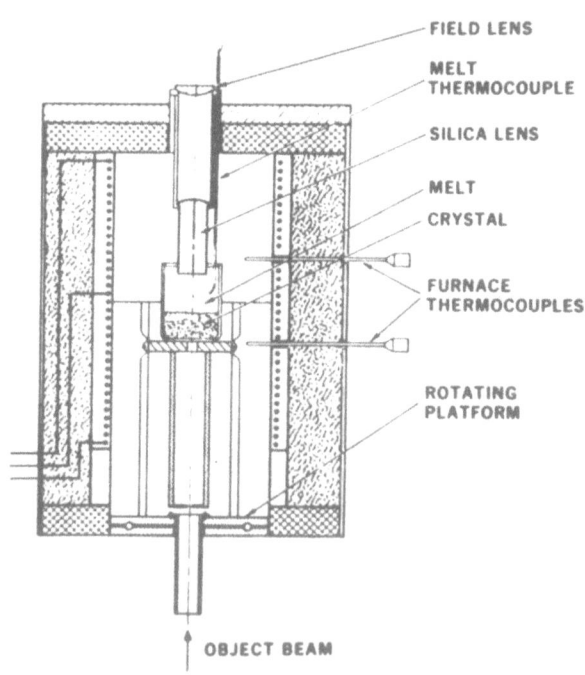

Fig. 5. Close up of crystal growing furnace used by McFee (Ref. 15). Note location of silica lens and field lens.

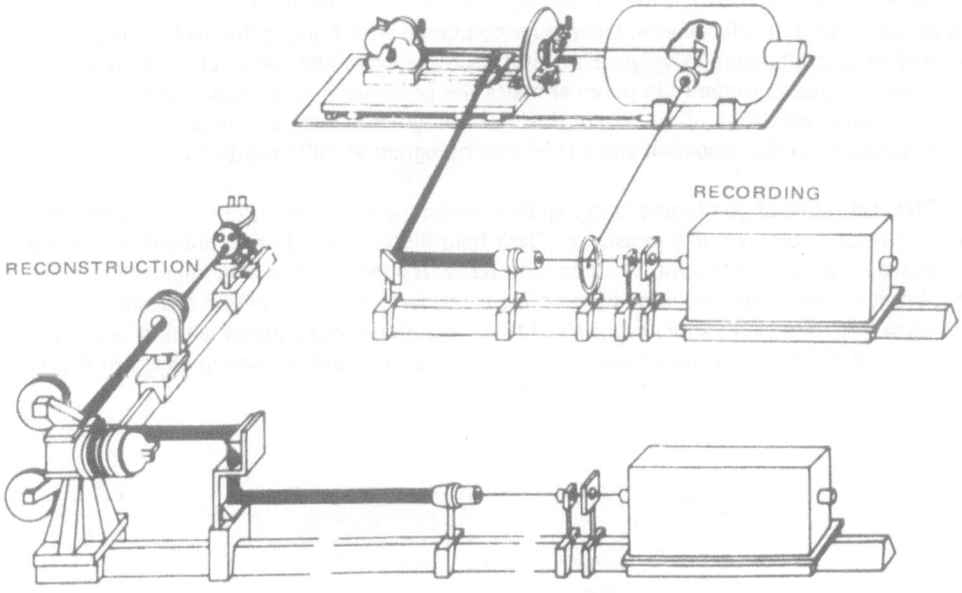

Fig. 6. Schematic diagram of holographic microscope developed by Cox, Buckles, and Whitlow (Ref. 16).

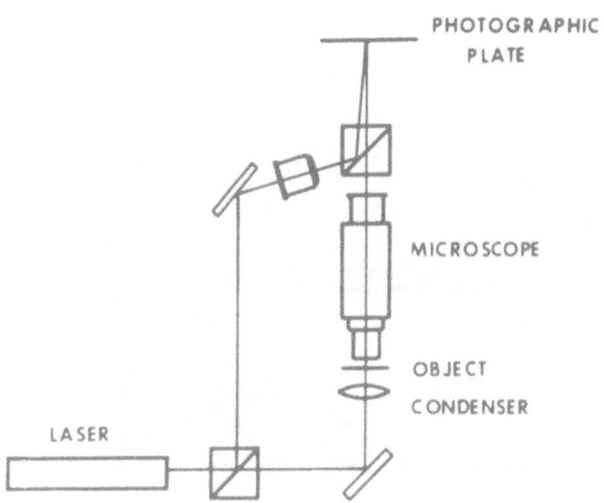

Fig. 7. Schematic diagram of holographic microscope developed by vanLigten and Oster-berg (Ref. 17).

Recently Rhodes has used this system to examine crystallization in polymer thin films(18). Using interferometric techniques combined with holographic techniques, secondary crystallization in polymer systems could be examined. Also, characterization of morphological boundaries in polymer melts was possible. Fig. 8 shows a portion of a polyethylene oxide spherulite using both single hologram reconstruction and simultaneous reconstruction of the hologram and a reference hologram at 300 magnifications.

This technique of combining holographic microscopy with interferometry stresses the versatility of a holographic microscope. Dark field illumination, phase-contrast, polarizing microscopy, all can be accomplished on a properly designed holographic microscope. Various image enhancement techniques, sampling procedures and information assessment processes can be used (19). Any and/or all of these techniques can be used without destroying the original hologram. Thus, several different techniques could be used to ascertain the information contained in the reconstructed image.

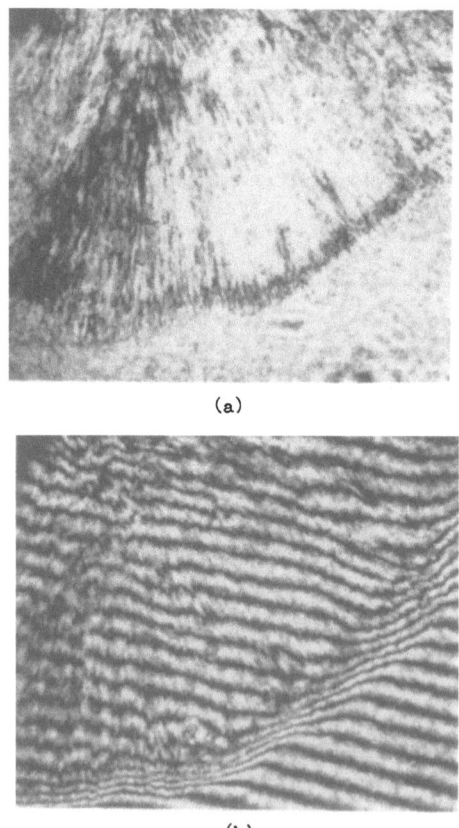

(a)

(b)

Fig. 8. Portion of a polyethylene oxide spherulite growing at 46°C. 300X magnification.
(a) Single hologram reconstruction (b) Simultaneous reconstruction of the field seen in
(a) with a reference hologram of the polymer melt.

CONCLUSION

The practical applications of holographic microscopy are many and varied. The design techniques are well-documented for a variety of problems. But holographic microscopy is not easy. The hardware is not packaged and optical components are not familiar to the laboratory technician. The alignment of the various components is delicate and time consuming, the recording medium must be handled carefully and images must be studied with great care.

There are situations where holographic microscopy is the only way to study the phenomena involved. In these situations, the care in design, construction, alignment, and use are rewarded. The image obtained upon reconstruction is truly a reward. The means one can use to study the image and the information gained from this study cannot be obtained in any other way. If a picture is worth a thousand words, then a hologram must be worth a thousand pictures.

REFERENCES

1. D. Gabor, *Proc. Roy. Soc. (London),* Ser. A, 197, 454 (1949).
2. W.L. Bragg, *Nature,* 149, 470 (1942).
3. D. Gabor, *Nature,* 161, 777 (1948).
4. E.N. Leith & J. Upatnieks, *J. Opt. Soc. Am.,* 52, 1123 (1962).
5. E.N. Leith & J. Upatnieks, *J. Opt. Soc. Am.,* 54, 1295 (1964).
6. We postpone discussion of the recording medium properties to the next section. For now, assume some light sensitive medium is available to record intensity variations.
7. Eastman Kodak and Agfa Gavaert make the best photographic films for holography. Consult the scientific applications division for advice.
8. Film manufacturers and suppliers can provide processing data. For other materials, consult the supplier.
9. E.N. Leith & J. Upatnieks, *J. Opt. Soc. Am.,* 54, 1295 (1964).
10. B.J. Thompson, *J. Soc. Phot. Inst. Engr.,* 4, 7 (1965).
11. H.J. Caulfield, *Opt. Comm.,* 3, 322 (July, 1971).
12. R.W. Meier, *J. Opt. Soc. Am.,* 56, 219 (1966).
13. R.F. vanLigten, and H. Osterberg, *Nature,* 211, 282 (1966).
14. L. Toth, and S.A. Collins, *Appl. Phys. Letters,* 13, 7 (1968).
15. R.H. McFee, *J. Opt. Soc. Am.,* 59, 474 (1969); 59, 1540 (1969); *Appl. Opt.,* 9, 1834 (1970).
16. M.E. Cox, R.G. Buckles, and D. Whitlow, *Appl. Opt.,* 10, 128 (1971).
17. R.F. vanLigten, and H. Osterberg, *Nature,* 211, 282 (1966).
18. M.B. Rhodes, *Appl. Opt.,* 13, 2263 (1974).
19. M.E. Cox, *Proceedings of the SPIE,* 52, 119 (1975).

BIBLIOGRAPHY

Born, M., and E. Wolf, *Principles of Optics,*Pergamon Press, (1969).
Cathey, W.T., *Optical Information Processing and Holography,* Wiley (1974).
Caulfield, H.J., and Sun Lu, *The Applications of Holography,* Wiley (1970).
DeVelis, J.B., and G.O. Reynolds, *Theory and Applications of Holography,* Addison-Wesley (1967).

Goodman, J.W., *Introduction to Fourier Optics,* McGraw-Hill (1968).
Smith, H., *Principles of Holography,* Wiley (1969).

ACHIEVING OPTIMUM OPTICAL PERFORMANCE THROUGH A

CUSTOMIZED MAINTENANCE PROGRAM

JERRY L. WOODBURY *

INTRODUCTION

Lack of brightness, poor clarity, unsatisfactory resolution, long exposure times, and time consuming microscopic analyses are just a few of the numerous problems that can arise from a dirty, poorly operating microscope or metallograph.

Achieving and maintaining optimum performance can be accomplished easily and with a minimum amount of expense by incorporating a maintenance program that has been specifically designed or "customized" for your specific laboratory.

Every metallographer is involved to some extent in the microscopy of metallic specimens. The degree of involvement varies with each individual position, but in most cases the security and/or personal satisfaction of this person's job depends on the results produced from his microscope or metallograph. Productivity, or lack of productivity, also is related to the condition of this instrument. It therefore becomes extremely important that this one instrument be kept in excellent operating order at all times.

Every metallographer and microscopist should think of his laboratory as somewhat unique, and therefore, should have a special "maintenance program" or plan to keep his optical instruments operating to their optimum performance level. If you have seen or experienced some of the problems outlined here, I believe you will agree as to the necessity of a "customized maintenance program' as presented on the following pages.

DEGRADATION OF A MICROSCOPE

Dust and dirt are found in every laboratory and are probably the biggest contributors to a poorly operating microscope or metallograph. Dust build-up on eyepieces, objectives, plane glass reflectors, first surface mirrors, prisms and other lenses within the optical system seriously reduce overall brightness, clarity, and resolution (Figs. 1 and 2). Lack of brightness is especially apparent on projection and viewing screens, and thereby reduces their overall effectiveness and value.

---

* Adolph I. Buehler Inc., Evanston, Illinois 60204 USA.

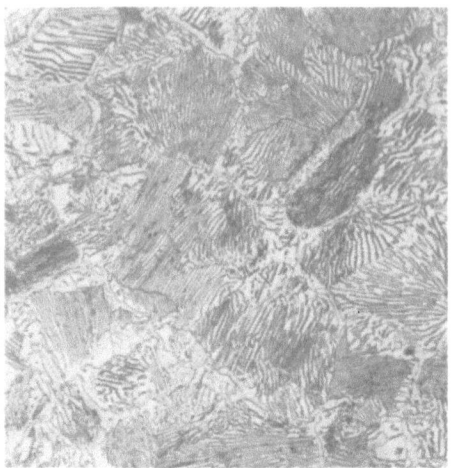 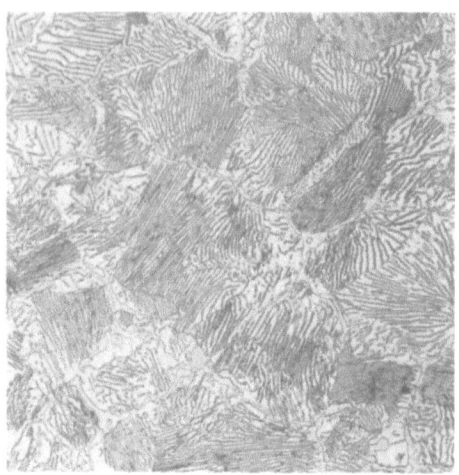

Fig. 1. Poorly resolved structure due to
finger smudge on objective lens. 500x
(Reduced 25% for reproduction)

Fig. 2. Clearly resolved structure after
proper cleaning of objective lens. 500 x
(Reduced 25% for reproduction)

Photomicrographs at high magnifications become increasingly difficult as exposure
times build up in direct proportion to dirt and dust build-up. Long exposure times in
most laboratories are detrimental, and in some laboratories where vibration is a major
problem, long exposure times make photomicrography impossible.

Lack of brightness also seriously affects observation in dark field, interference contrast, and polarized light where you generally need all the illumination your light source
can produce. This normally means opening diaphragms, turning intensity control knobs
to their maximum positions, removing density filters, re-positioning condensers; i.e.,
doing whatever has to be done to obtain additional illumination for his microscope or
metallograph. With an extremely dirty microscope you may never be able to provide
enough light to effectively use these important modes of illumination. Again, their overall value is either sharply decreased or eliminated altogether.

The degradation of a microscope's brightness, clarity, and resolution may also be due
to poor alignment of the illumination source (Fig. 3), cracked or shattered condensing
lenses, poorly aligned field and aperture diaphragms, improperly adjusted prisms or plane
glass reflectors, heat-crazed heat filters, improperly installed lenses, and a variety of
other causes.

Since this degradation often occurs slowly over a period of months or years, it is difficult for the operator to notice the change, and therefore, he slowly and continually in-
corporates compensations. For lack of brightness, he may open the field and/or aperture
diaphragms slightly, or turn the intensity control knob on the light source up a degree or
two, or remove a diffusion or possibly a green filter. *Each* of these compensations contributes to poorer resolution, clarity, and uneven illumination. The use of an illuminator

Fig. 3. Uneven illumination due to mis-aligned light source. 500x
(Reduced 25% for reproduction)

at maximum wattage sharply decreases bulb life, increases operating cost, and shortens
the length of time between microscope down-time.

A poorly operating optical system is both difficult to use and undesirable from a
laboratory manager's standpoint in that his technician or operator may very easily become frustrated and non-productive, simply because he is having extreme difficulty
in resolving detail in the image he sees (or barely sees.)

MECHANICAL TYPE PROBLEMS

Dust and dirt not only cause problems with optical components, but also seriously
affect smooth operation of the mechanical components. These problems often affect
the image in such a way as to appear to be an optical problem, and therefore produce
a detrimental effect to overall clarity and resolution. Typical problems of this nature
include drifting stages, wobbling or stiff stages, chipped or worn rack and pinions,
cracked or worn bellows, fine and/or coarse focus knobs stiff or frozen, objective
turret stop clips missing or worn, shutters sticking or otherwise working improperly,
and various other mechanical type problems. Often these problems develop from operator misuse such as setting very large, heavy specimens on the stage (Fig. 4) or using
wrenches to loosen tight focus knobs.

Although a few of the minor problems can be corrected by re-centering field and
aperture diaphragms, realigning light sources, replacement of bulbs with bad contacts
or deposits, and replacement of scratched or cracked mirrors and lenses; many, however, have progressed to a stage that replacement of defective parts is the only alternative. Some of this can be done as need arises by a skilled technician but many tasks

Fig. 4. Misuse of metallograph by placing large, heavy samples on stage.

must be performed by a factory trained serviceman who has the skill, proper tools and
the time. Periodic preventive maintenance *can* eliminate costly emergency visits, and
valuable down-time.   Since even the most conscientious microscopist can lose track of
time between service intervals, it would definitely be to his advantage to work closely
with his serviceman to schedule routine service visits.

A "CUSTOMIZED" MAINTENANCE PROGRAM

Each laboratory is somewhat unique in regard to cleanliness, vibration, temperature
fluctuations, personnel, etc., and therefore, requires maintenance more or less frequently,
to different degrees, and with special emphasis on certain potential problem areas.

For example, a laboratory which is in a high vibration location, such as a foundry or
stamping plant, will experience problems with heavy dust and grit build-up and mechanical problems caused by heavy, repeated vibrations (Table I).

A recommended *SERVICE PROGRAM* for this laboratory with its unique and quite
severe problems would have special emphasis on a very thorough cleaning and repair or
replacement of internal parts damaged by abnormally high vibration (Table II).

The *INITIAL INSPECTION* is performed by first discussing the condition of the
microscope with the main user to determine the nature and extent on the problems. This
is followed by an intensive visual check by the serviceman to locate any additional problems that the user is not aware of.

With this particular type of laboratory it is especially important that a very thorough
EXTERNAL CLEANING be performed. Normally, this should be performed with all
optics and accessories in place to avoid the possibility of introducing cleaning solutions

TABLE I. TYPICAL SERVICE PROBLEMS— FOUNDRY LABORATORY

I. **EXCESSIVE DUST AND GRIT**

   EFFECTS ON MICROSCOPE:
   A. Heavy Build-Up on Outer Surfaces of Optics and Cabinetry.
   B. Heavy Build-Up on Internal Lenses, Mirrors, Bulbs, etc.
   C. Accumulation of Grit on Shutter.
   D. Grit Contaminated Lubricants.

II. **ABNORMALLY HEAVY VIBRATIONS**

   EFFECTS ON MICROSCOPE:
   A. Loosening of Screws and Nuts.
   B. Loosening and Possible Damaging of Internal Optics.

TABLE II. RECOMMENDED SERVICE PROGRAM— FOUNDRY LABORATORY

A. Initial Inspection.

B. External Cleaning Prior to Disassembly.

C. Internal Cleaning; Dismantle, Clean, Parts Replacement, Lubrication.

D. Alignment of Light Source and Diaphragms.

E. Centering of Stage and Objectives.

F. Inspection/Cleaning of Desk and Accessories.

G. Final Performance Check.

H. Recommended Frequency of Service Visits: *Every 6 Months.*

and additional grit and dirt into the already dirty internal optical system (Fig. 5).

Following external cleaning, the metallograph should be completely disassembled and cleaned internally. This should include all objectives, eyepieces, mirrors, prisms, filters, diaphragms, plus any other lenses or mechanical components encountered (Fig. 6).

After the metallograph is cleaned and re-assembled, alignment of the light source, diaphragms and other lenses is logically the next step. This is a very critical service function, and it must be performed properly to assure both uniform and maximum illumination.

The *FINAL PERFORMANCE CHECK* is accomplished, with the user, by operating a variety of the mechanical components and also by examining a few highly polished

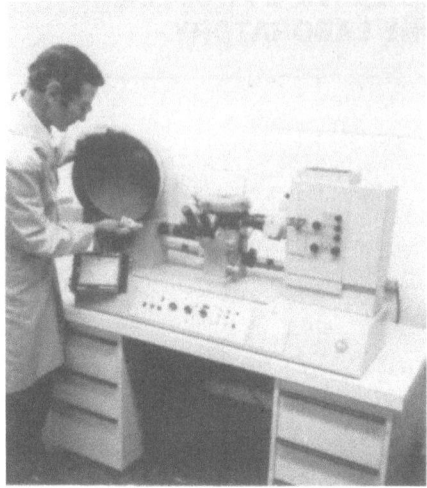

Fig. 5. External cleaning of metallograph.

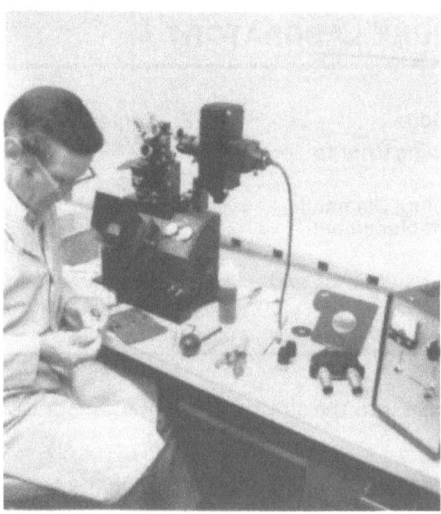

Fig. 6. Internal cleaning, parts replacement and lubrication.

specimens. Any overlooked areas or poorly cleaned optics will be noticed and can then easily be redone at this time. This is also a very appropriate time for the serviceman to point out to the user potential problem areas and minor preventative maintenance including the *RECOMMENDED FREQUENCY OF SERVICE*.

A plating, electronic or PC board laboratory will have to place special emphasis on rust and corrosion problems as well as mechanical problems that are the direct result of rust (Table III).

TABLE III. TYPICAL SERVICE PROBLEMS—
PC BOARD LABORATORY

I. CORROSIVE ATMOSPHERE

EFFECTS ON MICROSCOPE:

A. Surface Rust on External Metallic Surfaces.

B. Moderate/Severe Corrosion of Internal Parts.

C. Pitting/Clouding of Lenses and Mirrors.

D. Corrosion of Wires, Contacts and Connections.

E. Frozen Mechanical Components.

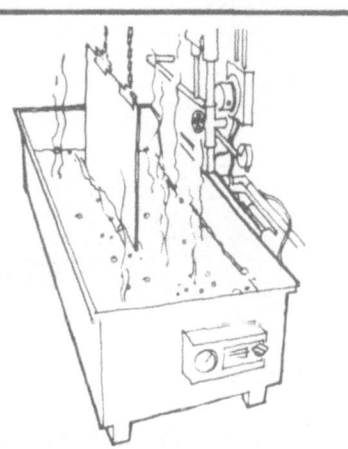

This very corrosive atmosphere not only affects the outer glass and metallic surfaces of a microscope, but it also can work uninterrupted on the internal optical and mechanical parts. Often, if a microscope has not been serviced on a frequent basis, this corrosion may have proceeded to the degree that some mechanisms become inoperative (Table IV).

TABLE IV. RECOMMENDED SERVICE PROGRAM—
PC BOARD LABORATORY

A. Initial Inspection.

B. External Cleaning and Rust Removal.

C. Internal Cleaning; Dismantle, Clean, Parts Replacement, Lubrication.

D. Alignment of Light Source and Diaphragms.

E. Inspection/Cleaning of Desk and Accessories.

F. Final Performance Check.

G. Recommended Frequency of Service Visits: *Every 6 Months.*

When a microscope reaches this condition, complete disassembly, cleaning, lubrication and application of a rust preventative wherever possible, is required (Fig. 7).

Special attention should also be given to pitted and cloudy lenses. Often this condition can be remedied by proper cleaning; however, if the anti-glare coating is scratched or removed, lens replacement will be the only alternative.

A microscope located in a particularly corrosive atmosphere is also likely to have electrical problems such as corroded wires and connections, sticking contacts and switches, and a variety of other problems of this nature. Special care should therefore be

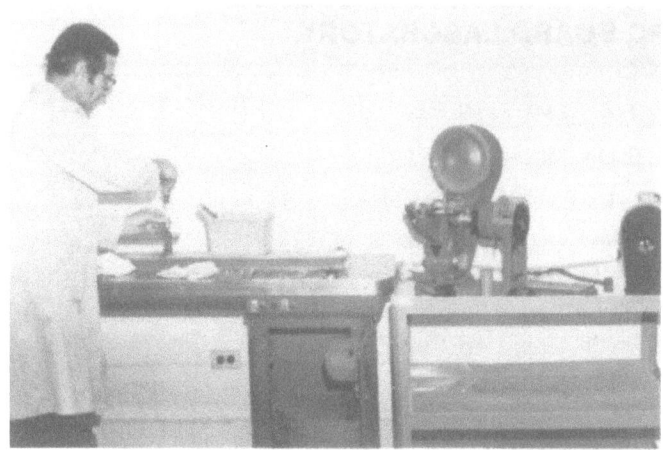

Fig. 7. Complete disassembly, cleaning and lubrication.

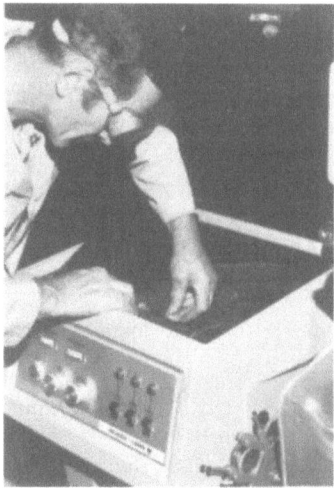

Fig. 8. Inspection of electrical wires and connections.

given to these often overlooked areas (Fig. 8) in an attempt to locate potential problem areas, make minor repairs on the spot and thus avoid major break-downs at a later date.

A school or university laboratory has an even different set of conditions under which it must operate. The *USE OF THE INSTRUMENT BY MANY PERSONS* is probably the main cause of problems in this type of laboratory. The operation of a metallograph by several skilled and unskilled persons typically results in problems such as misalignment of the light source, removed or broken filters, scratched lenses, etc. (Table V).

TABLE V. TYPICAL SERVICE PROBLEM— SCHOOL/UNIVERSITY LABORATORY

I. USE OF INSTRUMENT BY MANY PERSONS

EFFECTS ON MICROSCOPE:

A. Mis-Alignment of Light Source.
B. Removed or Damaged Filters.
C. Burned-Up Polarizers/Analyzers.
D. Hand Smudges on Lenses.
E. Decentered Objectives.
F. Broken Plane Glass Reflectors.
G. Lenses Installed Backwards.
H. Scratched Optical Surfaces.
I. Loose Electrical Connections.
J. Frozen Stage Locks.

A service program for this type of laboratory would include minor cleaning and lubrication as always, but with special emphasis on getting the microscope or metallograph back into normal operating condition (Table VI).

TABLE VI. RECOMMENDED SERVICE PROGRAM— SCHOOL/UNIVERSITY LABORATORY

A. Initial Inspection.

B. External Cleaning.

C. Internal Cleaning
   Optics
   Mechanical Parts
   Lubrication

D. Alignment of Microscope
   Light Source
   Diaphragms
   Mechanics

E. Final Performance Check.

F. Recommended Frequency of
   Service Visits: *Every 6-12 Months.*

The *INITIAL INSPECTION* step for this type of lab would in this case be performed with as many as possible of the professors or teachers who are continually using or are in charge of the microscope or metallograph. Again, the purpose of this is to determine the nature and extent of all the problems they have jointly experienced. A visual check by the serviceman would also be performed to determine if there are any additional problems.

Following internal cleaning, it is extremely important in this type of laboratory situ - ation, that all accessories be cleaned and examined to be sure they are in proper operating condition (Fig. 9).

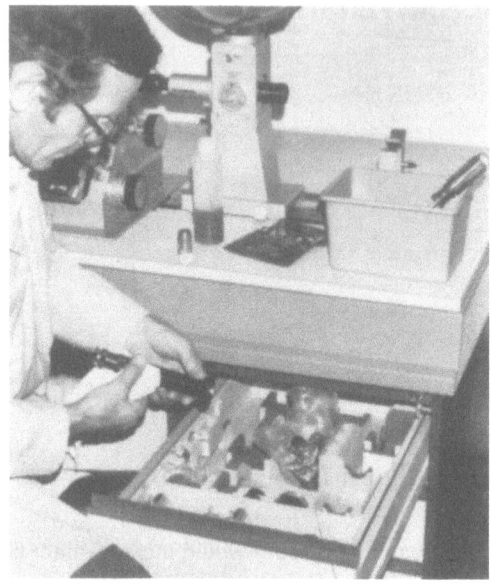

Fig. 9. Inspection and cleaning of desk and accessories.

There are some laboratories, of course, that do not have any of these 'special' conditions and can be classified as an average metallograph laboratory. Service visits to these labs are generally less frequent, less costly, and often require moderate cleaning and some alignment procedures (Tables VII and VIII).

TABLE VII. TYPICAL SERVICE PROBLEMS—
AVERAGE METALLOGRAPHIC LABORATORY

I. **MODERATE DUST AND GRIT ACCUMULATION**

    EFFECTS ON MICROSCOPE:
    A. Contaminated Optics.
    B. Tight Stage and Focusing Knobs.
    C. Sticking Camera Shutter.

II. **OCCASIONAL USE BY INEXPERIENCED OPERATOR**

    EFFECTS ON MICROSCOPE:
    A. Minor Mis-Alignment of Light.
    B. Incorrect Selection of Optics and Accessories.
    C. Mis-Adjustment of Controls.

TABLE VIII. RECOMMENDED SERVICE PROGRAM— AVERAGE METALLOGRAPHIC LABORATORY

A. Initial Inspection.

B. External Cleaning.

C. Internal Cleaning and Lubrication.

D. Alignment of Illumination System.

E. Final Performance Check.

F. Recommended Ferquency of Service Visits: *Every 12 Months.*

SUMMARY

Each metallographer can design his own optical maintenance program or programs for his lab, or he can have it done for him during a routine service visit. What is most important to remember, is that if you want the best possible results out of your microscope or metallograph, you must make it a point to have it serviced on a routine basis, by trained personnel, with *your* specific laboratory conditions in mind.

AUTHOR INDEX

A

Adams, M.D., 1—15
Agar, A.W., 128, 143
Allen, M.D., 49, 63
Anderson, K., 128, 143

B

Bainbridge, J.E., 164, 168
Batchelder, G.M., 119, 123, 125
Bayard, M., 141, 142
Beraha, E., 47, 62
Bohrer, W., 51, 63
Bomar, E.S., 47, 62
Born, M., 181
Brown, P.E., 153, 168
Buckles, B.G., 177, 179, 181

C

Cain, F.M., 151, 168
Campbell, J.J., 45, 62
Cathcart, J.V., 45, 62
Cathey, W.T., 181
Caulfield, H.J., 175, 181
Charsley, E.L., 139, 143
Cochran, F. L., 160, 168
Collins, S.A., 176, 181
Cook, W.H., 49, 63
Cox, M.E., 169—182
Crouse, R.S., 43—63

D

Dengel, D., 118, 125
Determan, H., 9, 14
Develis, J.B., 181
Dodd, J., 141
DuBose, C.K.H., 47, 62

E

Egorov, V.M., 118, 125
Emerson, W.G., 117, 125
Evans, J.H., 145—168
Exner, H.E., 48, 63

F

Fedotov, A.I., 118, 125
Fett, T., 118, 125
Finlay, G., 142
Flugge, S., 78

G

Gabler, F., 143
Gabor, D., 169, 170, 181
Gifkins, R.C., 26, 42
Goodman, J.W., 182
Gordon, A.M., 53, 63
Gwathmey, A.T., 45, 62
Gray, R.J., 18, 42, 43—63, 148, 161, 167, 168
Greaves, R.H., 17, 42
Greer—Spencer, J.G., 157, 168
Grube, W.L., 128, 143

H

Hacker, W.A., 136
Hartshorne, N.H., 135, 136, 143
Hays, C., 118, 125
Higgins, G.C., 10, 15
Hyzer, W.C., 6, 14

I

Iannone, B.N., 96

SUBJECT INDEX